MORE PRAISE FOR

A TIME TRAVELLER'S GUIDE TO LIFE, THE UNIVERSE, & EVERYTHING

'Every panel is packed with information and minute detail, and filled with humor that keeps the pace feeling speedy . . . It has the potential to open up physics and the realm of Einstein to unsuspecting comics fans, which means Flitcroft's journey is a success.'

– *Publishers Weekly*

'Dr. Flitcroft and Mr. Spencer have created a showcase for why comics and graphic novels can be the perfect teaching tools.'

– *New York Journal of Books*

'Reading [this book] is like taking an interstellar field trip with the best teacher in the world.'

– *Den of Geek*

'Communicates an enthusiasm to explore the cosmos with clarity, humor, and surprisingly subtle insights. This book will, I hope, inspire and enthuse a new generation to explore and to think.'

– Dr Andrew Pinsent, Ian Ramsey Centre for Science and Religion, Oxford University

*TO THE SUN AND PLANETS OF MY OWN LITTLE SOLAR SYSTEM:
JEAN, CALLUM, MYLES, AND OLIVER.*

TO THE FALLACY OF SUNK-COST.

First published 2015 by The O'Brien Press
12 Terenure Road East, Rathgar, Dublin 6
Tel: +353 1 492 3333; Fax: +353 1 492 2777
E-mail: books@obrien.ie. Website: www.obrien.ie
Originally published in 2013 by One Peace Books

ISBN: 978-1-84717-728-5

Copyright © 2013 by Ian Flitcroft
Illustrations copyright © 2013 by Britt Spencer

All rights reserved. No part of this publication may be reproduced or utilised in any form or by any means, electronic or mechanical, including photocopying, recording or in any information storage and retrieval system, without permission in writing from the publisher.

Cover design and illustrations by Britt Spencer.
Additional ink work by Allie Jachimowicz, Emily Spencer,
and Lomaho Kretzmann. Editing by Erin Canning.
Photo (page 75) copyright © 2007 iStockphoto LP/Jaap Hart

1 3 5 7 8 6 4 2
15 17 19 20 18 16

Printed and bound by GraphyCEMS, Spain
The paper in this book is produced using pulp from managed forests.

A TIME TRAVELLER'S GUIDE TO LIFE, THE UNIVERSE, & EVERYTHING

IAN FLITCROFT & BRITT SPENCER

THE O'BRIEN PRESS
DUBLIN

CONTENTS

SOME BIG QUESTIONS TO GET YOU STARTED...
(AND WHERE TO FIND THE ANSWERS)   6

CHAPTER 1: THE POWER OF THOUGHT   9

CHAPTER 2: LIGHT-YEARS, CHICKEN TAILS, AND ARMPITS   15

CHAPTER 3: STARS, ATOMS, AND LEGO®   21

CHAPTER 4: ALBERT EINSTEIN AND THE ATOM BOMB   28

CHAPTER 5: LET THERE BE LIGHT!   33

CHAPTER 6: THE BIG BANG AND THE MICROWAVE OVEN   38

CHAPTER 7: DARK MATTER AND HOW WIMPS MAY SAVE THE UNIVERSE   43

CHAPTER 8: THE STORY OF GRAVITY: THE WEAKLING, THE PLAGUE, AND THE APPLE   47

CHAPTER 9: ISAAC NEWTON'S LINK TO HARRY POTTER AND HOW TO WEIGH A PLANET   52

CHAPTER 10: EXPLODING STARS AND WHY EVERYONE HAS A LITTLE STAR QUALITY   57

CHAPTER 11: LITTLE GREEN MEN AND BLACK HOLES   62

CHAPTER 12: THE END OF THE DARK AGES AND THE INVENTION OF FROZEN CHICKEN   67

CHAPTER 13: A BRIEF HISTORY OF LIGHT   71

CHAPTER 14: WHY YOU SHOULD NEVER TRUST A QUANTUM MECHANIC   77

CHAPTER 15: QUANTUM MECHANICS FOR CAT LOVERS: ISAAC NEWTON STRIKES BACK   81

CHAPTER 16: CLOCKING THE SPEED OF THE FASTEST THING IN THE UNIVERSE   86

CHAPTER 17: RELATIVELY SIMPLE: AN EASY GUIDE TO RELATIVITY   91

CHAPTER 18: RELATIVELY WEIRD: HOW TO GROW YOUNGER, SHORTER, AND HEAVIER AT THE SAME TIME   96

CHAPTER 19: YES, BUT IS IT TRUE? ALBERT EINSTEIN'S THEORIES UNDER THE MICROSCOPE — 102

CHAPTER 20: LIFE IN THE GALACTIC SUBURBS AND WHY NEWS TAKES SO LONG TO GET THERE — 107

CHAPTER 21: DEAR ALIENS, MAY WE PRESENT PRESIDENT JIMMY CARTER — 112

CHAPTER 22: TO BE OR NOT TO BE? IS THAT A PLANET I SEE BEFORE ME? — 117

CHAPTER 23: IF I HAVE SEEN FURTHER: THE NOSE, THE MOOSE, AND THE TELESCOPE — 123

CHAPTER 24: ARE THERE PLANETS AROUND OTHER STARS? — 129

CHAPTER 25: THE OUTER, BIGGER PLANETS OF THE SOLAR SYSTEM — 134

CHAPTER 26: THE INNER, SMALLER PLANETS OF THE SOLAR SYSTEM — 141

CHAPTER 27: A CLOSE ENCOUNTER OF THE SOLAR KIND — 147

CHAPTER 28: WHAT IS LIFE AND WHERE DID IT COME FROM? — 153

CHAPTER 29: EATING SUNLIGHT: THE IMPORTANCE OF LIGHT FOR LIFE — 159

CHAPTER 30: ARE WE ALONE? LISTENING FOR EXTRATERRESTRIAL LIFE — 163

CHAPTER 31: WHY GREENHOUSE GASES ARE GOOD FOR THE PLANET — 168

CHAPTER 32: BLUE SKIES, RAINBOWS, AND THE MEANING OF LIFE — 173

CHAPTER 33: MORE BLACK HOLES AND WHY THE SPEED OF LIGHT ISN'T CONSTANT AFTER ALL — 178

CHAPTER 34: THE CURIOUS SENSATION OF BEING SEEN — 183

CHAPTER 35: WHAT HAPPENS IN YOUR BRAIN AND WHAT HAPPENED TO ALBERT EINSTEIN'S BRAIN — 189

CHAPTER 36: ALBERT EINSTEIN'S LOST SECRET REVEALED: THE ONE THING THAT CAN TRAVEL FASTER THAN LIGHT — 195

INDEX — 202

ACKNOWLEDGMENTS — 205

ABOUT THE AUTHOR AND ILLUSTRATOR — 207

SOME BIG QUESTIONS TO GET YOU STARTED...
(AND WHERE TO FIND THE ANSWERS)

HOW BIG IS THE GALAXY?
OF COURSE YOU KNOW IT'S REALLY BIG, BUT TURN TO PAGE 15 TO FIND OUT JUST HOW BIG. YOU'LL ALSO NEED TO KNOW ABOUT LIGHT-YEARS TO MAKE SENSE OF THE ANSWER. THESE ARE EXPLAINED ON THE SAME PAGE.

DOES THE UNIVERSE CONTAIN TRUTH AND BEAUTY?
THE ANSWER TO THIS QUESTION INVOLVES QUARKS AND IS FOUND ON PAGE 27.

HOW ABOUT $E=MC^2$?
YOU'VE HEARD OF IT, BUT WHAT DOES IT MEAN? ALL IS REVEALED ON PAGES 28-32, ALONG WITH THE STORY OF THE ATOMIC BOMB.

WHAT IS LIFE?
EASY QUESTION, HUH? WELL, TRY AND ANSWER IT YOURSELF, AND THEN HAVE A LOOK AT PAGES 153-154.

HOW DID LIFE START IN THE FIRST PLACE?
THIS SLIPPERY LITTLE QUESTION IS TACKLED ON PAGES 156-158.

WHY IS SPACE CALLED SPACE?
TO FIND OUT, TURN TO PAGE 107.

WHAT IS THE BIG BANG THEORY?
ONE CLUE: IT'S NOT A TV SERIES. TO DISCOVER WHAT THE BIG BANG IS REALLY ALL ABOUT AND HOW A CATHOLIC PRIEST INVENTED IT, FLIP TO PAGES 33-37.

WHAT DO MICROWAVE OVENS AND EXPLODING EGGS HAVE TO DO WITH THE UNIVERSE?
QUITE A LOT, SURPRISINGLY! FIND OUT HOW ON PAGES 41-42.

COULD WIMPS AND MACHOS DECIDE THE FATE OF THE UNIVERSE?
THEY JUST MIGHT. TO LEARN WHY, TAKE A LOOK AT PAGES 45-46.

WHY IS EVERY SINGLE PERSON ON THE PLANET ATTRACTIVE?
IT HAS TO DO WITH APPLES, TREES, AND A BRILLIANT BUT GRUMPY ENGLISHMAN WHO ALMOST DIED THE DAY HE WAS BORN. FOR A MORE USEFUL ANSWER, TRY PAGES 47-56.

WHAT IS A BLACK HOLE?
THE SCARY SPACE ONE IS EXPLAINED ON PAGES 63-66, BUT FOR A TOTALLY DIFFERENT TYPE OF BLACK HOLE, SEE PAGES 178-179.

WHICH FAMOUS ASTRONOMER WORE A FALSE NOSE AND HAD A PET MOOSE?
IF YOU DON'T BELIEVE ME, TURN TO PAGES 123-124.

HOW CAN LIGHT GO AROUND CORNERS?
REALLY! IT CAN. GO AND TAKE A LOOK AT PAGE 73.

IS THE SPEED OF LIGHT CONSTANT?
LIKE A LOT OF THINGS, IT DEPENDS ON WHERE YOU ARE. FIND OUT MORE ON PAGES 95-101.

WHAT IS THE DIFFERENCE BETWEEN SPECIAL AND GENERAL RELATIVITY?
WHAT DO YOU MEAN YOU NEVER KNEW THERE WERE DIFFERENT TYPES OF RELATIVITY? SHAME ON YOU! GO STRAIGHT TO PAGES 91–106.

HOW DO WE KNOW RELATIVITY IS REAL AND NOT JUST SOME CRAZY THEORY?
THAT'S WHAT SCIENTISTS ARE FOR! FOR A QUICK ANSWER, CHECK OUT PAGES 102–106.

DID SCHRÖDINGER HARM ANY CATS IN HIS WORK ON QUANTUM MECHANICS?
TO LEARN MORE ABOUT SCIENCE'S MOST FAMOUS FELINE, TAKE A LOOK AT PAGES 83–85.

WHERE DO SOME SCIENTISTS THINK IT RAINS DIAMONDS?
WHY WAIT TO WIN THE LOTTERY? YOU JUST NEED A SPACESHIP, LOTS OF TIME, AND A LARGE SACK. FOR DIRECTIONS, SEE PAGE 136.

DO ALIENS EXIST?
YOU COULD, OF COURSE, JUST ASK ONE, BUT IF YOU CAN'T FIND ONE NEARBY, READ PAGES 164–166.

WHY MIGHT PRESIDENT JIMMY CARTER BE THE FIRST HUMAN VOICE HEARD BY ALIENS?
HONESTLY, THIS ISN'T SCIENCE FICTION! FOR THE FACTS, TURN TO PAGES 115–116.

IF THE SKY IS BLUE, WHY ARE CLOUDS WHITE?
PEOPLE OFTEN ASK WHY THE SKY IS BLUE, BUT WHAT ABOUT CLOUDS? FIND OUT ON PAGE 175.

WHAT IS THE BIGGEST LIVING ORGANISM ON EARTH?
THINK YOU KNOW THIS ONE? IF WHAT YOU'RE THINKING ABOUT ISN'T THREE MILES (4.8 KM) ACROSS, YOU'RE WRONG. THE REAL ANSWER IS ON PAGE 156.

HOW DOES DNA STORE OUR GENETIC INFORMATION?
EVERYONE TALKS ABOUT DNA THESE DAYS, BUT DO YOU UNDERSTAND IT? IF NOT, TAKE A LOOK AT PAGES 154–156.

WHAT IS THE MEANING OF LIFE?
IF YOU DON'T ALREADY KNOW THE ANSWER, READ PAGES 176–177 TO SEE IF THAT HELPS.

DID ALBERT EINSTEIN HAVE A REALLY BIG BRAIN?
EINSTEIN WAS CERTAINLY VERY SMART, BUT TO FIND OUT THE SURPRISING STORY OF HIS BRAIN, FLIP TO PAGE 193.

WHY IS SEEING SO MUCH HARDER THAN IT LOOKS?
THE ANSWER, MY FRIEND, IS BURIED INSIDE YOUR HEAD. IF YOU DON'T WANT TO OPEN UP YOUR SKULL TO FIND OUT, TRY PAGES 189–193 INSTEAD.

HOW CAN BLACK AND WHITE LOOK THE SAME?
TO FIND OUT WHY SEEING ISN'T ALWAYS BELIEVING, SEE PAGE 179.

WHAT DOES TRAVEL FASTER THAN LIGHT (FINALLY PROVING EINSTEIN WRONG)?
THERE REALLY IS ONE THING, AND YOU'LL HAVE TO READ THE BOOK TO FIND OUT... SORRY, I CAN'T GIVE AWAY ALL OF THE ANSWERS!

CHAPTER 1: THE POWER OF THOUGHT

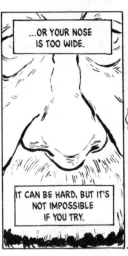

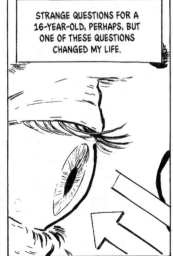

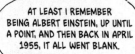

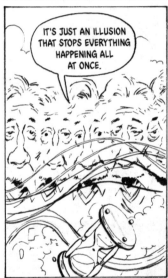

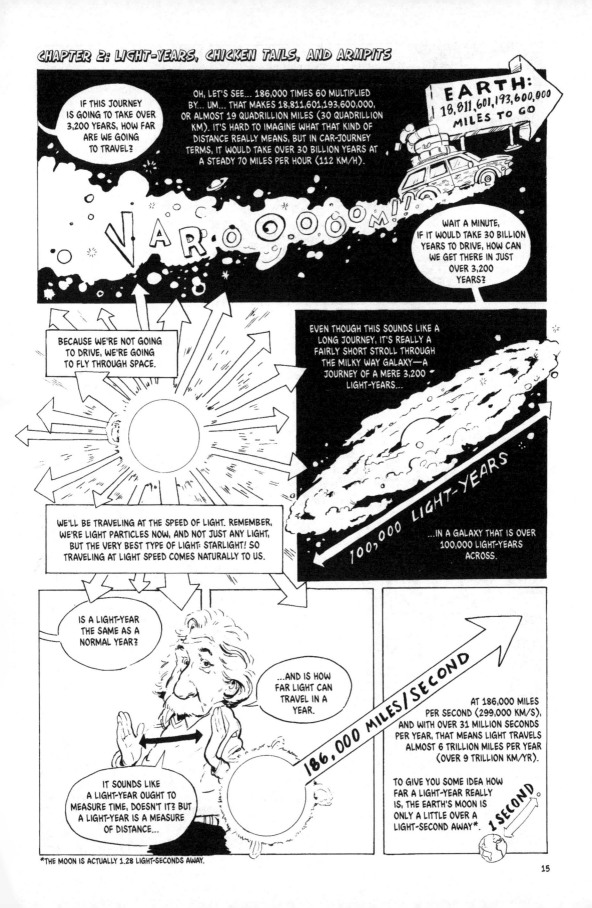

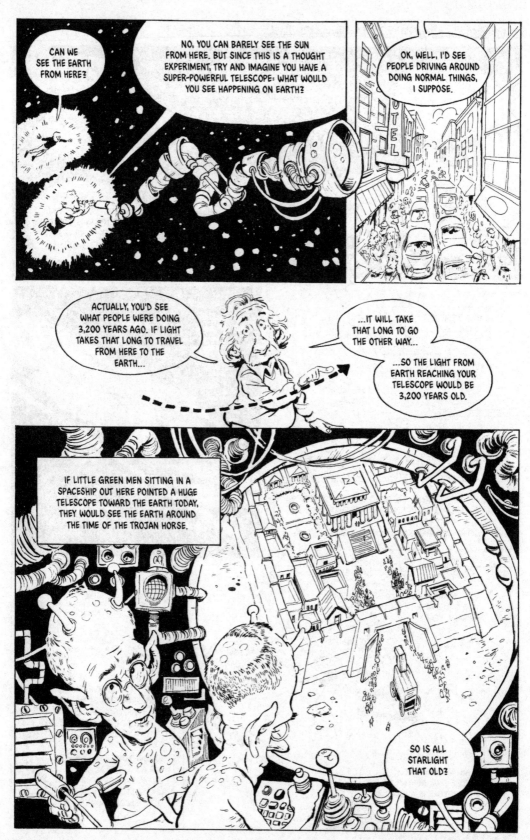

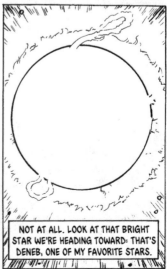

*AS LONG AS YOU ARE AT LEAST 7 YEARS OLD. THERE ARE ALSO TWO STARS FOR 4-YEAR-OLDS AND ONE FOR 5-YEAR-OLDS, BUT NONE FOR 6-YEAR-OLDS.

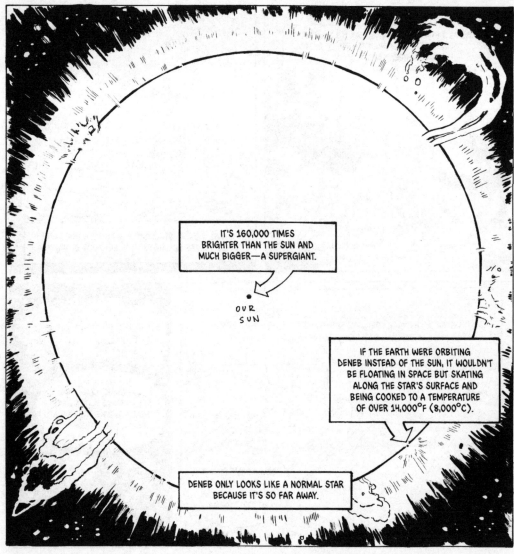

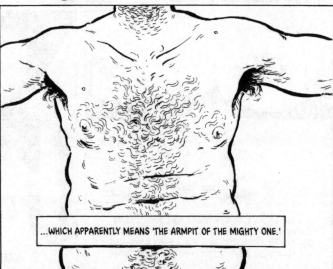

CHAPTER 3: STARS, ATOMS, AND LEGO.

That's what the German scientist Julius Robert von Mayer tried to work out. He calculated that without some source of energy, the sun could only shine for around 5,000 years.

Not for scientists. Mind you, some people thought the earth wasn't much older. An Irish archbishop, James Ussher, calculated that the world began on October 23, 4004 BC.

THIS WONDERFULLY SIMPLE EQUATION SAYS THAT A LITTLE MATTER TURNED INTO PURE ENERGY RELEASES A LOT OF ENERGY.

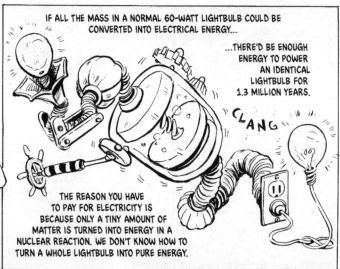

IF ALL THE MASS IN A NORMAL 60-WATT LIGHTBULB COULD BE CONVERTED INTO ELECTRICAL ENERGY...

...THERE'D BE ENOUGH ENERGY TO POWER AN IDENTICAL LIGHTBULB FOR 1.3 MILLION YEARS.

THE REASON YOU HAVE TO PAY FOR ELECTRICITY IS BECAUSE ONLY A TINY AMOUNT OF MATTER IS TURNED INTO ENERGY IN A NUCLEAR REACTION. WE DON'T KNOW HOW TO TURN A WHOLE LIGHTBULB INTO PURE ENERGY.

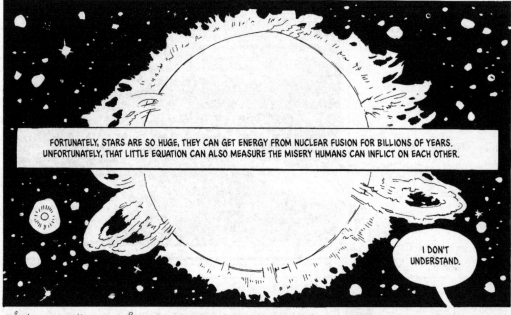

FORTUNATELY, STARS ARE SO HUGE, THEY CAN GET ENERGY FROM NUCLEAR FUSION FOR BILLIONS OF YEARS. UNFORTUNATELY, THAT LITTLE EQUATION CAN ALSO MEASURE THE MISERY HUMANS CAN INFLICT ON EACH OTHER.

I DON'T UNDERSTAND.

WARS AND BOMBS. I TOLD YOU THAT STARS ARE LIKE CONTINUOUSLY EXPLODING ATOMIC BOMBS.

YEP.

WELL, THAT SAME EQUATION EXPLAINS WHY ATOMIC BOMBS ARE SO DESTRUCTIVE. IN SENSIBLE HANDS, ATOMIC ENERGY CAN BE A GREAT BENEFIT FOR HUMANITY...

...BUT SADLY, HUMANS ARE NOT ALWAYS SENSIBLE.

THE EQUATION? NO. I REGRETTED WRITING THE LETTER TO ROOSEVELT, BUT WHO KNOWS WHAT WOULD HAVE HAPPENED IF I HADN'T.

I THINK ROOSEVELT UNDERSTOOD THE SIGNIFICANCE OF THE ATOMIC BOMB. WHEN ROOSEVELT DIED, HARRY S. TRUMAN BECAME PRESIDENT AND SEEMED TO THINK OF THE IT AS JUST ANOTHER BOMB, ONLY BIGGER AND BETTER. HE SAW IT AS A FAST WAY TO END THE WAR WITH JAPAN.

I JUST HOPE THE WORLD HAS LEARNED THAT LESSON, THOUGH I'M SURE THE THE POINT COULD HAVE BEEN MADE WITHOUT SO MANY INNOCENT PEOPLE HAVING TO DIE.

AS TO THE EQUATION $E=MC^2$, MATHEMATICS DON'T HURT ANYONE, PEOPLE DO THAT. I ONLY WROTE IT DOWN. IT HAS EXISTED SINCE THE UNIVERSE BEGAN, AND WITHOUT IT, THERE WOULD BE NO STARS OR LIVING CREATURES.

STARS CAN SHINE FOR BILLIONS OF YEARS, ENOUGH TIME FOR THE FIRST SPARK OF LIFE TO FORM, GROW INTO MICROSCOPIC BUGS, THEN BIG BUGS, ALL THE WAY TO DINOSAURS, AND FINALLY US!

SO ARE THE SCIENTISTS WHO ARGUED ABOUT THE EARTH'S AGE SATISFIED?

AS SOON AS SCIENTISTS SOLVE ONE PROBLEM...

ALRIGHT I THINK I GOT IT!

...THERE IS ALWAYS ANOTHER ONE TO SOLVE, AND WE'VE BARELY SCRATCHED THE SURFACE OF SCIENCE YET.

WE'RE ONLY STARTING THE JOURNEY NOW?

THIS IS JUST THE BEGINNING.

WHERE ARE WE GOING?

OUR DESTINATION: A RATHER PRETTY BLUE-GREEN CIRCLE SURROUNDING A POOL OF BLACKNESS THAT IS SITUATED IN THE BACK GARDEN OF A HOUSE ON A PLANET CALLED EARTH. BUT THERE IS A LOT OF SPACE AND TIME BETWEEN HERE AND THERE.

CHAPTER 5: LET THERE BE LIGHT!

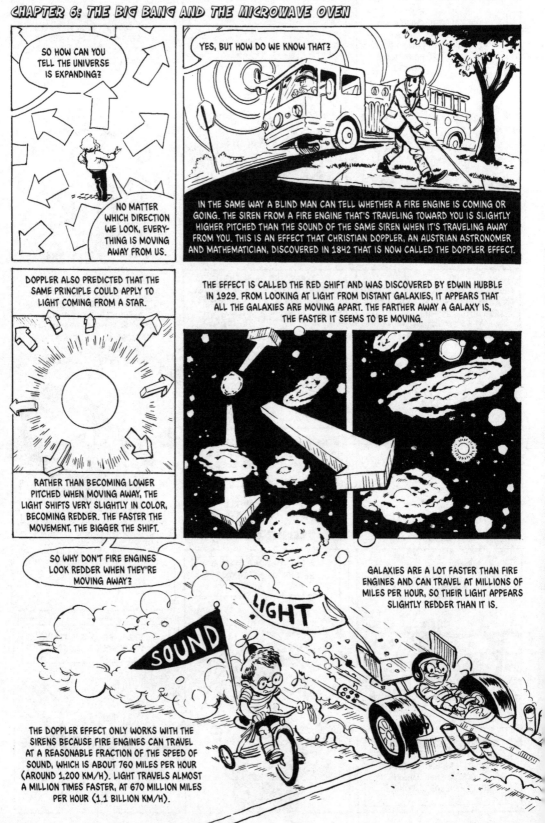

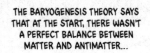

LIGHT CAN'T GET THROUGH A PIECE OF CHICKEN...
LIGHT
RADIO
...BUT A RADIO WAVE CAN PASS STRAIGHT THROUGH A CHICKEN AND CARRY ON HALFWAY AROUND THE WORLD.
MICROWAVES FALL SOMEWHERE IN THE MIDDLE: THEY CAN GET MOST OF THE WAY THROUGH A CHICKEN, BUT A LOT OF THEM WILL BE ABSORBED.
MICRO WAVE
WHEN ABSORBED, THEY GIVE UP THEIR ENERGY AND HEAT UP FOOD FROM THE INSIDE.
THE FACT THAT MICROWAVES CAN HEAT THINGS UP WAS ACCIDENTALLY DISCOVERED BY PERCY SPENCER, AN AMERICAN INVENTOR, IN 1946...
MAGNETRON
...WHEN HE WAS EXPERIMENTING WITH SOMETHING CALLED A MAGNETRON.
THE FIRST ODD THING HE NOTICED WAS THAT A CHOCOLATE BAR IN HIS POCKET MELTED WHEN HE WAS STANDING NEXT TO THE MAGNETRON.
HE THEN TRIED POPCORN, AND IT STARTED POPPING.
LATER THAT DAY, HE SHOWED HIS DISCOVERY TO SOMEONE ELSE IN THE LABORATORY, USING AN EGG.
AS THEY WATCHED THE EGG, IT EXPLODED ALL OVER THEIR FACES. IT WAS LIKE RECREATING THE BIRTH OF THE UNIVERSE.
EXPLODING AN EGG IN A MICROWAVE IS LIKE THE START OF THE UNIVERSE?
GEORGES-HENRI LEMAÎTRE ORIGINALLY CALLED THE BLOB OF MATTER THAT EXPLODED INTO THE UNIVERSE THE COSMIC EGG. DON'T YOU SEE? IT'S ALL CONNECTED: LEMAÎTRE'S COSMIC EGG, EXPLOSIONS, MICROWAVES, AND ME GETTING EGG ON MY FACE FOR NOT BELIEVING IN IT.
HMM, I DON'T KNOW.
WELL, NEXT TIME YOU MICROWAVE POPCORN, YOU SHOULD THINK ABOUT THE START OF THE UNIVERSE. BUT DON'T TRY THE EGG TRICK, JUST TAKE MY WORD FOR IT... IT EXPLODES.

CHAPTER 7: DARK MATTER AND HOW WIMPS MAY SAVE THE UNIVERSE

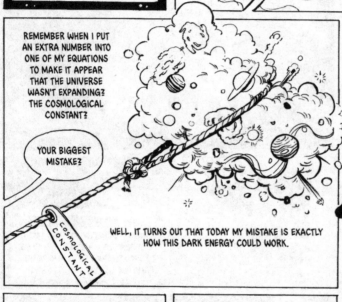

"SO WE'LL NEVER KNOW UNTIL WE START TRAVELING AROUND THE GALAXY."

NO, ASTRONOMERS HAVE A WAY TO FIND INVISIBLE THINGS; ODDLY ENOUGH, IT WASN'T INVENTED BY AN ASTRONOMER BUT BY KARL JANSKY, AN ENGINEER AT THE AMERICAN TELEPHONE COMPANY BELL LABS, IN 1931. JANSKY WAS LOOKING FOR ALL THE DIFFERENT TYPES OF INTERFERENCE, OR NOISE, THAT COULD AFFECT RADIO TELEPHONE SIGNALS. HE NOTICED THAT ONE TYPE OF INTERFERENCE APPEARED AROUND THE SAME TIME EACH DAY AND CAME FROM CERTAIN POSITIONS IN THE SKY—FROM OUTER SPACE. THIS DISCOVERY WAS THE BIRTH OF RADIO ASTRONOMY, OR THE MAPPING AND UNDERSTANDING OF RADIO SIGNALS THAT COME FROM STARS AND GALAXIES.

"ARE THESE LIKE ALIEN RADIO STATIONS?"

NO, RADIO WAVES CAN BE CREATED TO SEND MUSIC AND NEWS, BUT THE RADIO WAVES FROM SPACE SEEM TO BE NATURALLY OCCURRING, COMING FROM STARS, GALAXIES, AND OTHER STRANGE OBJECTS THAT LIVE IN SPACE. AND GUESS WHAT YOU CAN DETECT WITH A RADIO TELESCOPE?

"STARS?"

WELL, SOME STARS, BUT THE MOST IMPORTANT THINGS ARE THOSE THAT YOU CAN'T SEE WITH A REGULAR TELESCOPE, LIKE DUST AND INTERSTELLAR GAS.

"SO ASTRONOMERS FINALLY HAVE A WAY OF LOOKING FOR THE MISSING DARK MATTER."

EXACTLY, BUT THEY STILL DON'T THINK THEY'VE FOUND ALL OF IT. WIMPS MIGHT ALSO MAKE UP DARK MATTER, AND SOME SCIENTISTS THINK THEY WILL DETERMINE THE FATE OF THE UNIVERSE.

"THE FATE OF THE UNIVERSE LIES IN THE HANDS OF WIMPS? ALBERT, YOU REALLY HAVE LOST IT NOW."

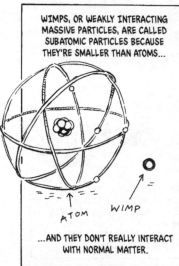

HE WAS TINY AND FEEBLE WHEN HE WAS BORN—SO SMALL, HIS MOTHER, HANNAH NEWTON, SAID HE'D FIT "INTO A QUART POT." NO ONE THOUGHT HE'D SURVIVE. FORTUNATELY, FOR SCIENCE'S SAKE, YOUNG NEWTON PROVED EVERYONE WRONG, LIVING FOR 84 YEARS, WHICH WAS ALMOST A MIRACLE BACK THEN.

HE ALSO SURVIVED AN OUTBREAK OF A DEADLY PLAGUE IN 1665 THAT CLOSED DOWN CAMBRIDGE UNIVERSITY, AND WAS SENT BACK TO THE LINCOLNSHIRE COUNTRYSIDE FOR SOME UNINTERRUPTED THINKING TIME.

CLOSED DUE TO PLAGUE

THIS WAS WHEN HE WAS SUPPOSEDLY SITTING UNDER THE APPLE TREE.

SURELY PEOPLE HAD ALWAYS KNOWN APPLES FELL OFF TREES?

OF COURSE THEY DID. BUT NEWTON ASKED THE QUESTION WHY? HE WAS ALSO SMART ENOUGH TO ANSWER THIS QUESTION BY DEVELOPING HIS THEORY OF GRAVITY.

WHAT WAS HIS THEORY?

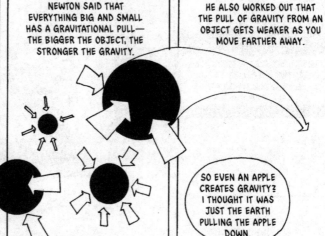

NEWTON SAID THAT EVERYTHING BIG AND SMALL HAS A GRAVITATIONAL PULL— THE BIGGER THE OBJECT, THE STRONGER THE GRAVITY.

HE ALSO WORKED OUT THAT THE PULL OF GRAVITY FROM AN OBJECT GETS WEAKER AS YOU MOVE FARTHER AWAY.

SO EVEN AN APPLE CREATES GRAVITY? I THOUGHT IT WAS JUST THE EARTH PULLING THE APPLE DOWN.

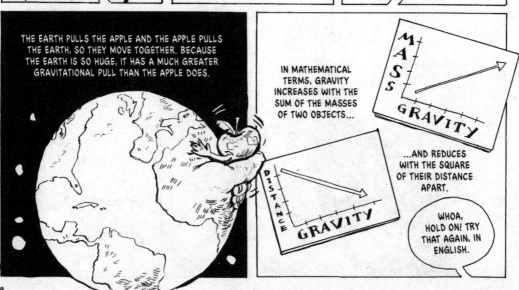

THE EARTH PULLS THE APPLE AND THE APPLE PULLS THE EARTH, SO THEY MOVE TOGETHER. BECAUSE THE EARTH IS SO HUGE, IT HAS A MUCH GREATER GRAVITATIONAL PULL THAN THE APPLE DOES.

IN MATHEMATICAL TERMS, GRAVITY INCREASES WITH THE SUM OF THE MASSES OF TWO OBJECTS...

...AND REDUCES WITH THE SQUARE OF THEIR DISTANCE APART.

WHOA, HOLD ON! TRY THAT AGAIN, IN ENGLISH.

CHAPTER 9:
ISAAC NEWTON'S LINK TO HARRY POTTER AND HOW TO WEIGH A PLANET

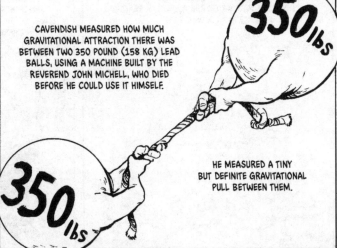

THINK OF A STAR AS A BALLOON: THE STRETCHY RUBBER OF THE BALLOON IS LIKE GRAVITY AND TRYING TO SHRINK THE STAR; THE PRESSURE OF THE AIR INSIDE THE BALLOON IS LIKE THE NUCLEAR REACTIONS AND TRYING TO MAKE THE BALLOON BIGGER. WHEN THEY ARE BALANCED, THE BALLOON STAYS THE SAME SIZE.

UNTIL THE AIR IN THE BALLOON ESCAPES?

THE ESCAPING AIR IS LIKE NUCLEAR FUSION STOPPING, SO THE STAR BEGINS TO CONTRACT. THIS HAPPENS WHEN THE CENTER RUNS OUT OF HYDROGEN FUEL AND THE NUCLEAR REACTIONS START TO SLOW DOWN.

THEN WHAT HAPPENS?

THE STAR CONTINUES TO CONTRACT AND THE SQUEEZING ACTION OF GRAVITY HEATS UP THE OUTER LAYERS THAT ARE STILL FULL OF HYDROGEN.

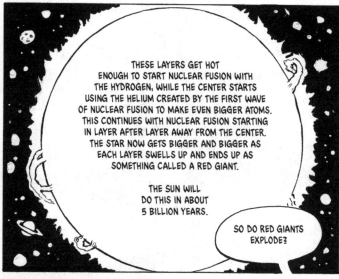

THESE LAYERS GET HOT ENOUGH TO START NUCLEAR FUSION WITH THE HYDROGEN, WHILE THE CENTER STARTS USING THE HELIUM CREATED BY THE FIRST WAVE OF NUCLEAR FUSION TO MAKE EVEN BIGGER ATOMS. THIS CONTINUES WITH NUCLEAR FUSION STARTING IN LAYER AFTER LAYER AWAY FROM THE CENTER. THE STAR NOW GETS BIGGER AND BIGGER AS EACH LAYER SWELLS UP AND ENDS UP AS SOMETHING CALLED A RED GIANT.

THE SUN WILL DO THIS IN ABOUT 5 BILLION YEARS.

SO DO RED GIANTS EXPLODE?

WHEN THEY CAN'T MAKE ENOUGH ENERGY FROM NUCLEAR REACTIONS TO COUNTER GRAVITY, THEY DO, BUT ONLY IF THEY'RE BIG ENOUGH.

WHEN THERE'S NOT ENOUGH NUCLEAR FUSION TO COUNTERACT THE PULL OF GRAVITY, STARS COLLAPSE IN ON THEMSELVES. A SMALL STAR LIKE THE SUN WON'T EXPLODE, IT WILL KEEP SHRINKING UNTIL IT'S ABOUT THE SAME SIZE AS THE EARTH. THESE TINY, OLD STARS ARE CALLED WHITE DWARFS.

EVERYTHING IS SO COMPRESSED IN A WHITE DWARF THAT A PIECE THE SIZE OF A SUGAR LUMP COULD WEIGH MORE THAN A TON.

"IRON DOESN'T EXPLODE, DOES IT?"

NOT BY ITSELF. IT'S WHEN A BIG STAR USES IRON ATOMS IN FUSION REACTIONS THAT EVERYTHING STARTS TO GO VERY WRONG. RATHER THAN RELEASE ENERGY, NUCLEAR FUSION REACTIONS WITH IRON ABSORB ENERGY.

INSTEAD OF PUSHING AGAINST GRAVITY, THE CENTER OF THE STAR SUDDENLY STARTS TO COLLAPSE EVEN MORE, UNTIL IT EXPLODES AS A SUPERNOVA.

IN THE LAST FEW DAYS BEFORE THE STAR EXPLODES, FUSION REACTIONS WITH IRON MAKE BIGGER AND BIGGER ATOMS, FROM GOLD AND SILVER TO FANCY ATOMS LIKE MOLYBDENUM.

REMEMBER HOW I TOLD YOU THAT HUMANS ARE MOSTLY MADE UP OF SIX TYPES OF ATOMS: CARBON, NITROGEN, OXYGEN, HYDROGEN, CALCIUM, AND PHOSPHORUS. WELL, ALL OF THESE ATOMS, EXCEPT HYDROGEN, WERE MADE INSIDE A STAR BEFORE IT DIED. THERE ARE SOME EVEN BIGGER ATOMS, LIKE MOLYBDENUM, IN HUMANS THAT MUST HAVE BEEN MADE IN A SUPERNOVA EXPLOSION MILLIONS OR BILLIONS OF YEARS BEFORE THE SUN FORMED.

"HOW DO THESE ATOMS GET INTO HUMANS?"

FROM THE GAS AND DUST THAT MAKE UP THE NEXT GENERATION OF STARS AND PLANETS. THAT'S WHAT THE SUN AND EARTH WERE BUILT FROM... AND SO ARE YOU.

"SO EARTH AND ALL THE PEOPLE ON IT ARE RECYCLED SPACE JUNK?"

EXACTLY, THERE'S A LITTLE BIT OF STAR QUALITY IN EVERY ONE OF US! MIND YOU, DON'T GET TOO PLEASED WITH YOURSELF. BUGS, EARTHWORMS, AND THE INK IN A PEN ALSO CONTAIN ATOMS THAT WERE ONCE IN A STAR.

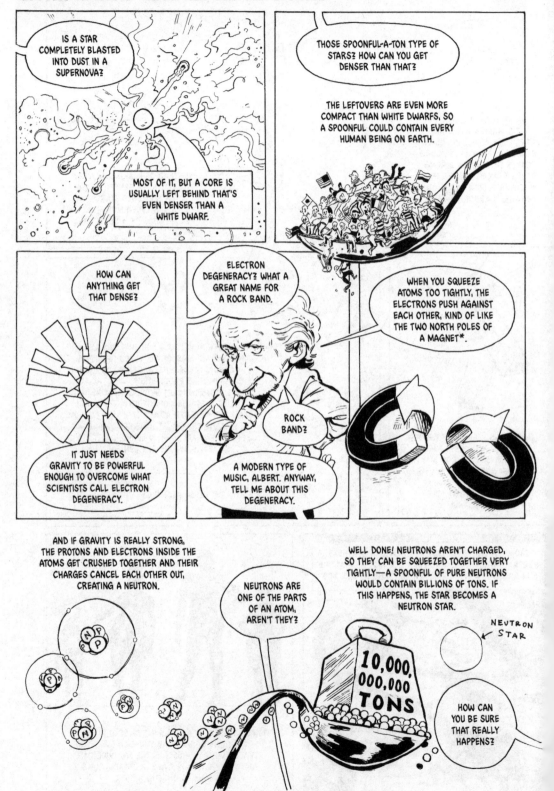

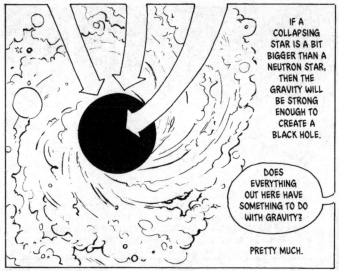

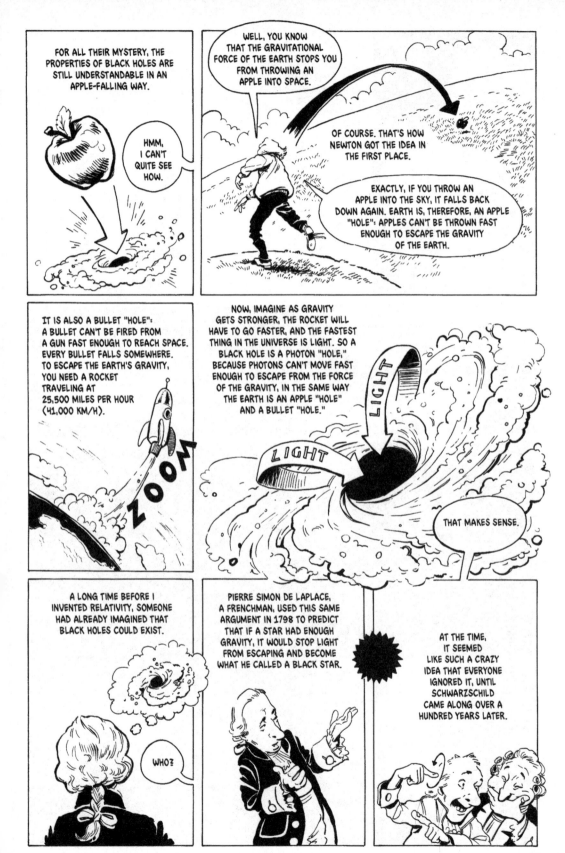

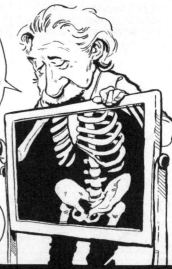

CHAPTER 12:
THE END OF THE DARK AGES AND THE INVENTION OF FROZEN CHICKEN

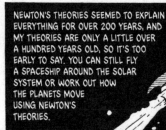

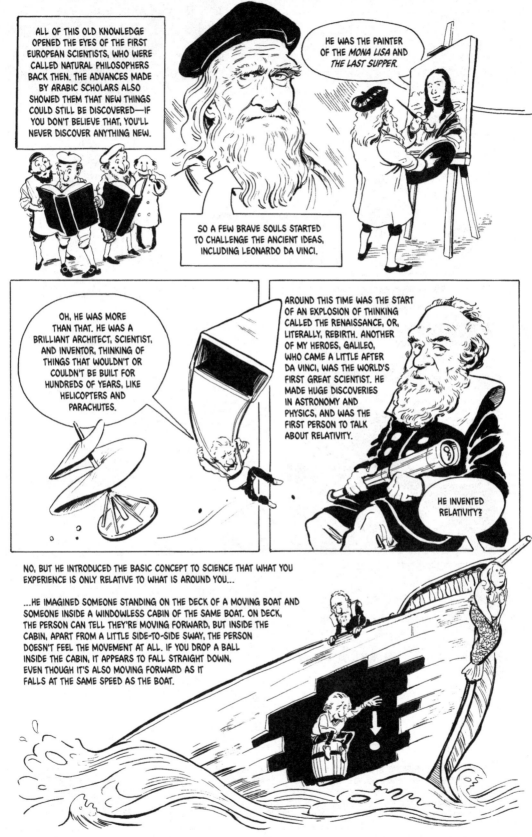

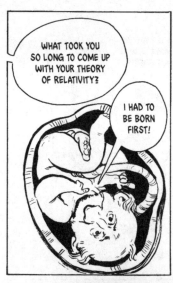

WHAT TOOK YOU SO LONG TO COME UP WITH YOUR THEORY OF RELATIVITY?

I HAD TO BE BORN FIRST!

IN THE MEANTIME, EUROPEANS WERE STILL BUSY WORKING OUT WHAT SCIENCE WAS. IN 1605, SCIENCE WAS GIVEN A KICKSTART WHEN AN ENGLISHMAN, FRANCIS BACON, PUBLISHED A BOOK CALLED *OF THE PROFICIENCE AND ADVANCEMENT OF LEARNING, DIVINE AND HUMAN.*

THIS BOOK GAVE BIRTH TO WHAT WE NOW THINK OF AS SCIENCE. RATHER THAN ONLY STUDYING THE WRITINGS OF THE GREAT GREEK PHILOSOPHERS, BACON URGED PEOPLE TO THINK FOR THEMSELVES AND COME UP WITH NEW THEORIES FOR HOW THE UNIVERSE WORKS.

BACON IS CONSIDERED THE FATHER OF MODERN SCIENCE.

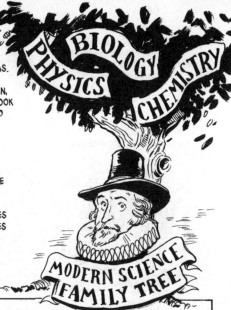

WHAT DID HE DISCOVER?

WELL, BACON INVENTED THE SCIENTIFIC PROCESS, OR AT LEAST A VERSION OF IT. HE THOUGHT THAT A THEORY WOULD NATURALLY COME FROM EXAMINING THE WORLD. IN HIS WRITINGS, HE DIDN'T TALK MUCH ABOUT EXPERIMENTS AND SPARKS OF CREATIVITY, BUT DURING HIS LAST WEEK ALIVE, HE MANAGED TO INVENT THE FROZEN CHICKEN... AND DIE AS A RESULT OF IT.

FRANCIS BACON DIED TRYING TO CREATE THE FIRST FROZEN CHICKEN?

ON A SNOWY MARCH DAY IN 1626, BACON WAS VISITING LONDON WITH THE KING OF ENGLAND'S DOCTOR WHEN HE HAD THE SPARK OF AN IDEA THAT COLD COULD STOP MEAT FROM ROTTING—THERE WERE NO REFRIGERATORS BACK THEN, OF COURSE. SO HE GOT OUT OF HIS WARM CARRIAGE IN HIGHGATE TO BUY A CHICKEN AND STUFF IT WITH SNOW.

SADLY, HE CAUGHT PNEUMONIA AND DIED A FEW DAYS LATER. BUT AT LEAST HIS CHICKEN STAYED FRESH UNTIL THEN.

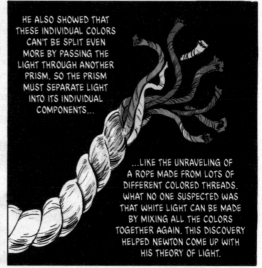

NEWTON BELIEVED LIGHT WAS MADE UP OF LITTLE PARTICLES, OR CORPUSCLES, AS HE CALLED THEM, OF DIFFERENT TYPES, EACH ONE REPRESENTING A DIFFERENT COLOR ACROSS THE RAINBOW, OR SPECTRUM, AS SCIENTISTS CALL IT, FROM RED TO VIOLET. WHITE LIGHT IS AN EQUAL MIXTURE OF ALL THE DIFFERENT TYPES OF LIGHT PARTICLES. EVERY HUE IMAGINABLE CAN BE MADE BY MIXING THE DIFFERENT COLORED TYPES IN VARYING PROPORTIONS.

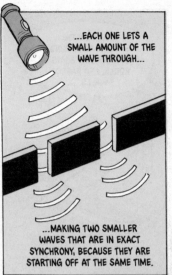

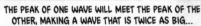

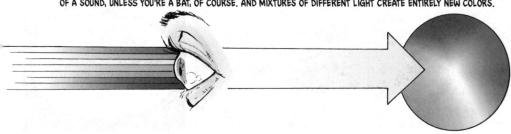

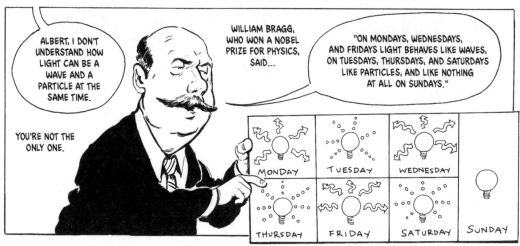

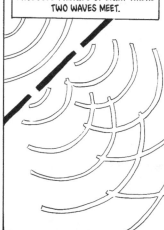

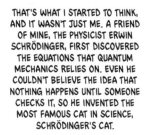

THAT'S WHAT I STARTED TO THINK, AND IT WASN'T JUST ME. A FRIEND OF MINE, THE PHYSICIST ERWIN SCHRÖDINGER, FIRST DISCOVERED THE EQUATIONS THAT QUANTUM MECHANICS RELIES ON. EVEN HE COULDN'T BELIEVE THE IDEA THAT NOTHING HAPPENS UNTIL SOMEONE CHECKS IT, SO HE INVENTED THE MOST FAMOUS CAT IN SCIENCE, SCHRÖDINGER'S CAT.

MEOW?

IF NOTHING HAPPENS UNTIL IT IS OBSERVED, THEN IMAGINE THE FOLLOWING SCENARIO: A CAT IS PLACED IN A BOX WITH A SMALL GADGET THAT WILL RELEASE POISON.

A REAL CAT?

NO, THIS IS AN IMAGINARY CAT, SO WHATEVER HAPPENS, A CAT ISN'T HARMED. LIKE OUR JOURNEY, IT'S A THOUGHT EXPERIMENT.

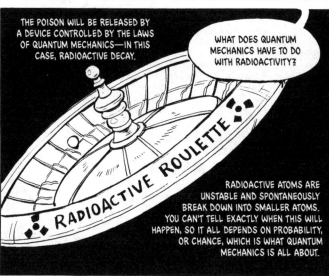

THE POISON WILL BE RELEASED BY A DEVICE CONTROLLED BY THE LAWS OF QUANTUM MECHANICS—IN THIS CASE, RADIOACTIVE DECAY.

WHAT DOES QUANTUM MECHANICS HAVE TO DO WITH RADIOACTIVITY?

RADIOACTIVE ATOMS ARE UNSTABLE AND SPONTANEOUSLY BREAK DOWN INTO SMALLER ATOMS. YOU CAN'T TELL EXACTLY WHEN THIS WILL HAPPEN, SO IT ALL DEPENDS ON PROBABILITY, OR CHANCE, WHICH IS WHAT QUANTUM MECHANICS IS ALL ABOUT.

IMAGINE A LUMP OF RADIOACTIVE MATERIAL AND A DEVICE TO DETECT IF AN ATOM HAS BROKEN DOWN.

...UNTIL SOMEONE OBSERVES THE RESULT.

CAN'T THE CAT TELL IF IT'S DEAD OR NOT?

ONLY IF IT'S ALIVE.

THAT'S CRAZY.

LET'S SAY THIS ATOMIC BREAKUP HAS A 50:50 CHANCE OF HAPPENING IN ONE HOUR, AND WHEN IT DOES, THE POISON IS RELEASED. UNTIL THE BOX IS OPENED AN HOUR LATER, BOTH OUTCOMES SHOULD COEXIST. ACCORDING TO QUANTUM MECHANICS, THE CAT SHOULD BE BOTH DEAD AND ALIVE...

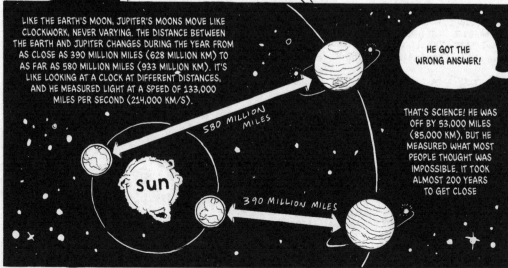

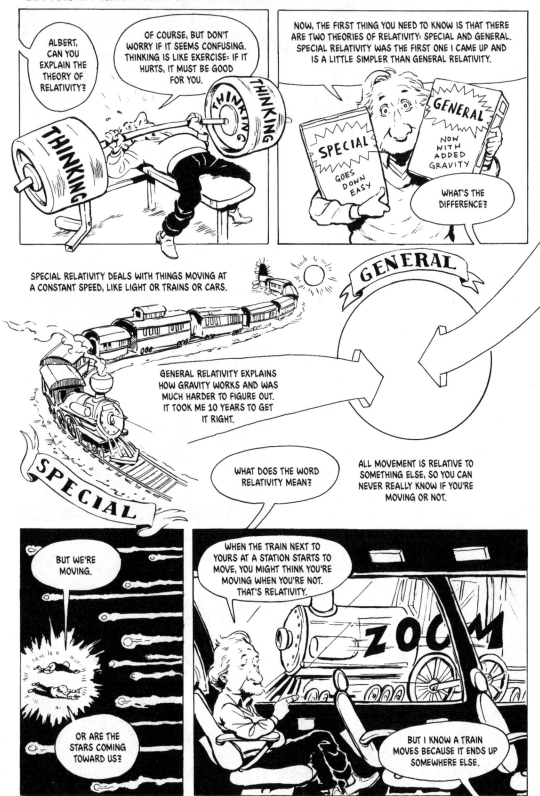

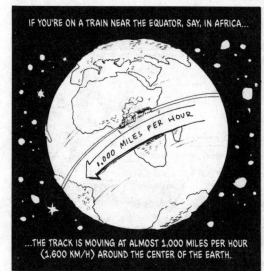

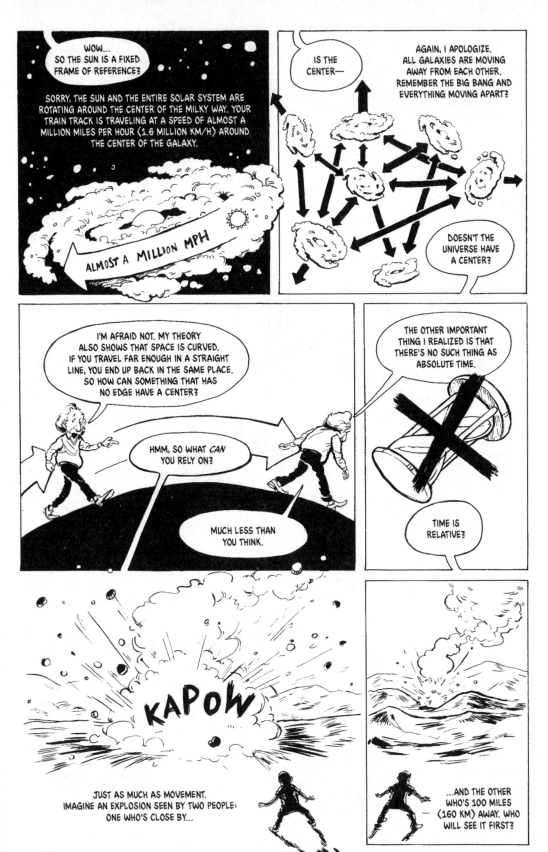

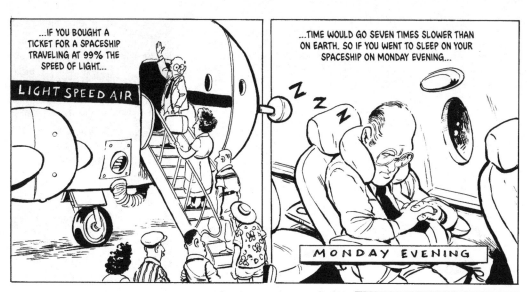

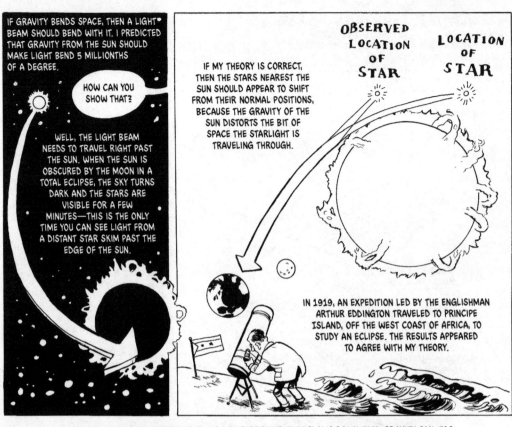

CHAPTER 20:
LIFE IN THE GALACTIC SUBURBS AND WHY NEWS TAKES SO LONG TO GET THERE

WHEN ANYTHING BIG HAPPENS, THE POSSIBLE CONSEQUENCES START SPREADING LIKE RIPPLES THROUGH SPACE AT THE SPEED OF LIGHT. UNTIL LIGHT, OR ANY SIGNAL, FROM THAT EVENT REACHES A CERTAIN POINT IN TIME AND SPACE, IT'S AS IF IT NEVER HAPPENED.

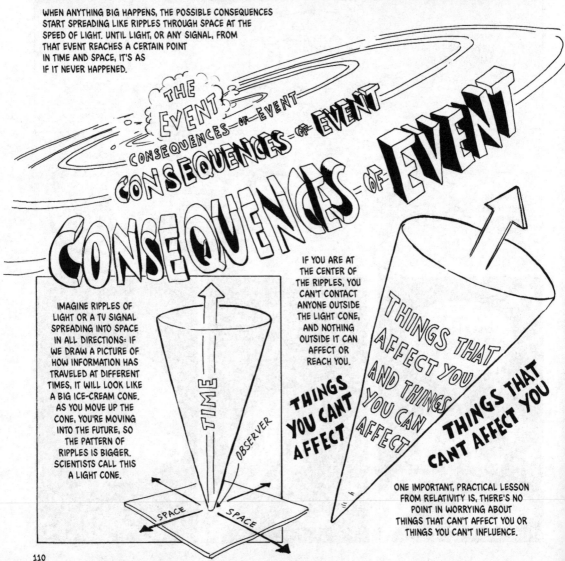

IMAGINE RIPPLES OF LIGHT OR A TV SIGNAL SPREADING INTO SPACE IN ALL DIRECTIONS: IF WE DRAW A PICTURE OF HOW INFORMATION HAS TRAVELED AT DIFFERENT TIMES, IT WILL LOOK LIKE A BIG ICE-CREAM CONE. AS YOU MOVE UP THE CONE, YOU'RE MOVING INTO THE FUTURE, SO THE PATTERN OF RIPPLES IS BIGGER. SCIENTISTS CALL THIS A LIGHT CONE.

IF YOU ARE AT THE CENTER OF THE RIPPLES, YOU CAN'T CONTACT ANYONE OUTSIDE THE LIGHT CONE, AND NOTHING OUTSIDE IT CAN AFFECT OR REACH YOU.

ONE IMPORTANT, PRACTICAL LESSON FROM RELATIVITY IS, THERE'S NO POINT IN WORRYING ABOUT THINGS THAT CAN'T AFFECT YOU OR THINGS YOU CAN'T INFLUENCE.

CHAPTER 22: TO BE OR NOT TO BE? IS THAT A PLANET I SEE BEFORE ME?

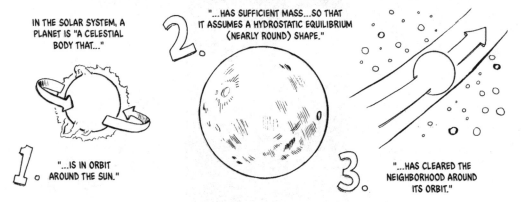

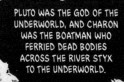

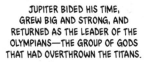

GALILEO ALSO DISCOVERED THAT THERE ARE FAR MORE STARS THAN CAN BE SEEN WITH THE NAKED EYE AND THAT THE MILKY WAY IS, IN FACT, MADE UP OF THOUSANDS OF STARS TOO FAINT TO BE SEEN WITHOUT A TELESCOPE.

HE ALSO LEARNED THAT JUPITER HAS FOUR MOONS REVOLVING AROUND IT.

WHY WAS THAT IMPORTANT?

That's him. Kepler worked out three simple rules, his laws of planetary motion, to explain how the planets move...

Kepler's Laws of Planetary Motion:

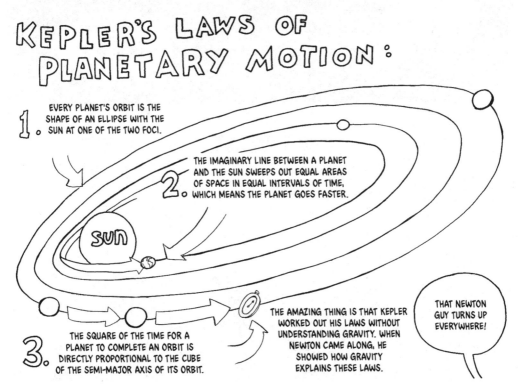

1. Every planet's orbit is the shape of an ellipse with the sun at one of the two foci.

2. The imaginary line between a planet and the sun sweeps out equal areas of space in equal intervals of time, which means the planet goes faster.

3. The square of the time for a planet to complete an orbit is directly proportional to the cube of the semi-major axis of its orbit.

The amazing thing is that Kepler worked out his laws without understanding gravity. When Newton came along, he showed how gravity explains these laws.

That Newton guy turns up everywhere!

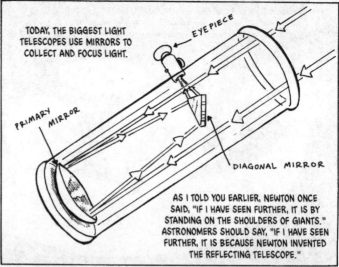

CHAPTER 24: ARE THERE PLANETS AROUND OTHER STARS?

IMAGINE YOU'RE LOOKING AT THE LARGE AND TINY SKATERS FROM A DISTANCE: YOU MIGHT ONLY BE ABLE TO SEE THE LARGE ONE, BUT YOU'LL BE ABLE TO TELL THERE'S A SECOND SKATER BECAUSE THE LARGE ONE WILL BE WOBBLING SLIGHTLY.

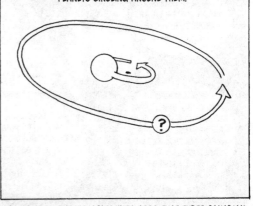

SO ASTRONOMERS LOOKED FOR STARS THAT APPEARED TO BE WOBBLING FROM SIDE TO SIDE IN SPACE BECAUSE OF THE GRAVITATIONAL PULL OF SMALL BUT INVISIBLE PLANETS CIRCLING AROUND THEM.

SOME PEOPLE CLAIMED TO HAVE FOUND WOBBLING STARS AS LONG AGO AS 1855, BUT IT WASN'T UNTIL 1988 THAT THREE CANADIAN ASTRONOMERS, BRUCE CAMPBELL, G. A. H. WALKER, AND S. YANG, FOUND THE FIRST STAR WITH A DEFINITE PLANET. EVEN THEN IT TOOK UNTIL 2003 FOR SCIENTISTS TO CONFIRM IT REALLY WAS A PLANET.

WILL I BE ABLE TO SEE THIS PLANET-WOBBLING?

YOU CAN TRY. THE PLANET IS GOING AROUND A STAR CALLED ALRAI, OR GAMMA CEPHEI, IN THE CONSTELLATION CEPHEUS.

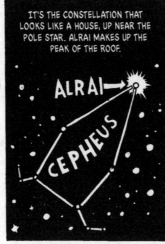

IT'S THE CONSTELLATION THAT LOOKS LIKE A HOUSE, UP NEAR THE POLE STAR. ALRAI MAKES UP THE PEAK OF THE ROOF.

ALRAI → CEPHEUS

ALRAI IS ONE OF THE BRIGHTEST STARS NEAR THE POLE STAR, SO IN AUTUMN AND WINTER IN THE NORTHERN HEMISPHERE, IT'S VISIBLE. BUT THE WOBBLE IS SO MINUTE, IT CAN ONLY BE DETECTED IN PHOTOGRAPHS TAKEN BY THE LARGEST TELESCOPES ON EARTH.

IS THIS THE ONLY PLANET THAT'S BEEN DISCOVERED SO FAR?

NOT AT ALL. ONCE THEY FOUND THE FIRST ONE, OTHER ASTRONOMERS STARTED LOOKING MORE SERIOUSLY.

I JUST LOOKED UP THE LIST, AND THERE ARE CURRENTLY 760 CONFIRMED PLANETS AROUND 609 STARS, AND A FEW THOUSAND OTHER POSSIBILITIES. AND NEW PLANETS ARE BEING FOUND ALL THE TIME...

...SO BY NEXT MONTH, THAT NUMBER WILL ALREADY BE OUT OF DATE.

REMEMBER JOHANNES KEPLER?

THE MAN WHO WORKED OUT HOW THE PLANETS MOVE?

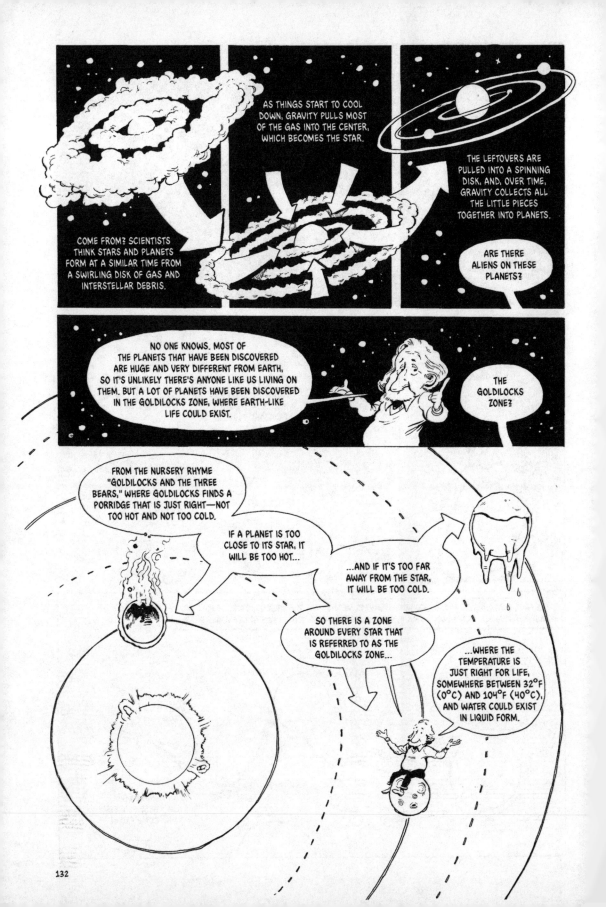

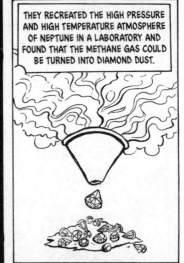
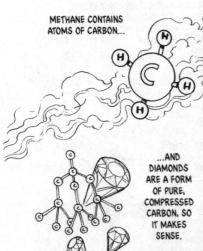

AN ICEBERG CAN BE VERY HEAVY, BUT AS LONG IT WEIGHS LESS THAN AN EQUAL VOLUME OF WATER, IT WILL FLOAT.

ICEBERG

EQUAL VOLUME OF WATER

ARCHIMEDES, ANOTHER OF THE ANCIENT GREEKS AND ONE OF THEIR BEST MATHEMATICIANS, WORKED ALL THIS OUT IN 212 BC, IN WHAT BECAME KNOWN AS THE ARCHIMEDES' PRINCIPLE.

HOW COULD HE HAVE WORKED OUT SATURN WOULD FLOAT? I THOUGHT THE ANCIENT GREEKS DIDN'T KNOW WHAT THE PLANETS REALLY WERE?

EUREKA!

THAT'S TRUE, BUT HE WORKED OUT THE PRINCIPLE THAT APPLIES TO ALL FLOATING OR SINKING THINGS—A STICK, A BOAT, OR A PLANET. IT ALL STARTED WHEN HE WAS ASKED TO FIGURE OUT IF A GOLDSMITH HAD CHEATED KING HIERON II OF SYRACUSE, WHEN MAKING A CROWN FOR HIM. WHEN THE IDEA CAME TO ARCHIMEDES IN THE BATH, HE SUPPOSEDLY RAN NAKED DOWN THE STREET SHOUTING, "EUREKA!"

ONCE THE CHEATING GOLDSMITH HAD BEEN DEALT WITH, ARCHIMEDES EXPANDED THE IDEA TO EXPLAIN HOW THINGS FLOAT. UNLIKE SOME OTHER THEORIES DATING FROM ANCIENT GREECE, THE ARCHIMEDES' PRINCIPLE HAS STOOD THE TEST OF TIME AND IS AS VALID TODAY AS WHEN ARCHIMEDES LEAPT OUT OF HIS BATH DRIPPING WITH WATER AND ENTHUSIASM.

THAT'S JUPITER, THE BIGGEST PLANET IN THE SOLAR SYSTEM.

WOW, THAT IS ONE MASSIVE PLANET OVER THERE!

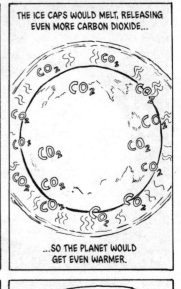

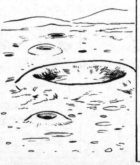

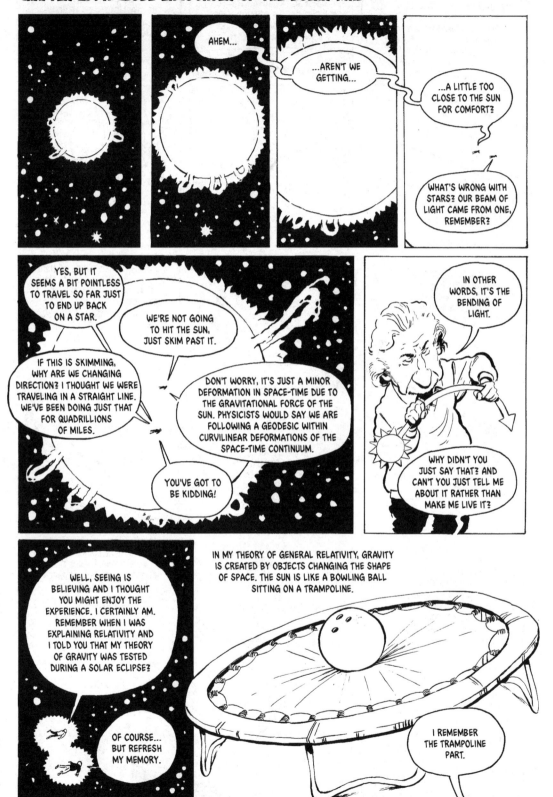

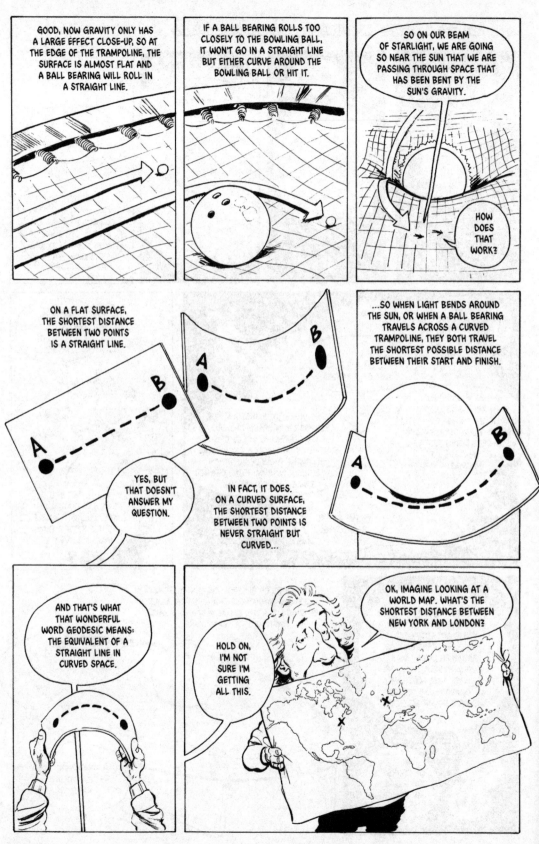

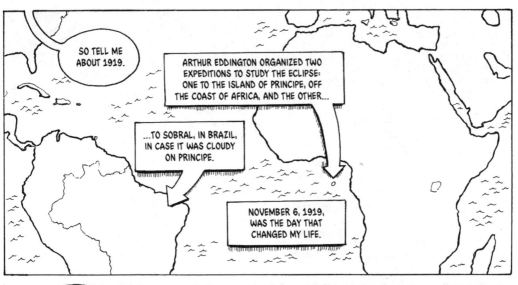

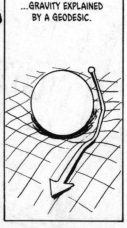

CHAPTER 28: WHAT IS LIFE AND WHERE DID IT COME FROM?

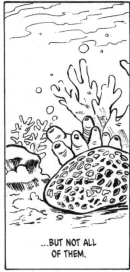

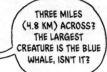

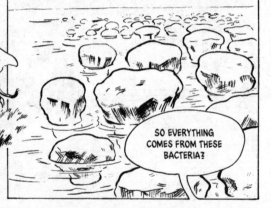

BEFORE THEM, THERE MUST HAVE BEEN EVEN SIMPLER FORMS OF LIFE, AND OVER BILLIONS OF YEARS, LIFE DEVELOPED INTO ALL THE FORMS WE SEE TODAY. BUT CHARLES DARWIN DIDN'T EVEN KNOW ABOUT DNA. HE DEVELOPED HIS THEORY OF EVOLUTION BY EXAMINING THE ANIMALS ON THE GALAPAGOS ISLANDS, A TINY SET OF ISLANDS IN THE PACIFIC OCEAN. HE FOUND THAT FINCHES LIVING ON DIFFERENT ISLANDS HAD CHANGED IN ORDER TO ACCLIMATE TO EATING PARTICULAR TYPES OF FOOD. DARWIN SUGGESTED THAT IF YOU STRING TOGETHER LOTS OF THESE LITTLE CHANGES OVER MILLIONS OF YEARS, YOU CAN EXPLAIN HOW ANY LIVING CREATURE COULD EVOLVE FROM A SIMPLE COMMON ANCESTOR.

CHAPTER 29: EATING SUNLIGHT: THE IMPORTANCE OF LIGHT FOR LIFE

HAVE YOU EVER WONDERED WHAT WOULD HAPPEN IF THE SUN STOPPED SHINING?

I CAN'T SAY THAT I HAVE. I SUPPOSE IT WOULD BE JUST LIKE NIGHTTIME.

FOR A LITTLE WHILE, BUT IMAGINE A NEVER-ENDING NIGHT: IT WOULD BE THE END OF NEARLY EVERY LIVING THING ON EARTH.

NEARLY?

THERE ARE BACTERIA AROUND VENTS IN THE OCEAN FLOOR—WHERE HOT VOLCANIC WATER AND GAS BUBBLE UP—THAT LIVE OFF THE HYDROGEN SULFIDE COMING OUT OF THESE VENTS. SO THEY ARE LIVING OFF CHEMICAL ENERGY, NOT LIGHT ENERGY, AND STRANGE GIANT WORMS LIVE OFF THESE BUGS.

MIND YOU, AS HYDROGEN SULFIDE HAS THE SMELL OF ROTTEN EGGS AND IS JUST AS POISONOUS AS HYDROGEN CYANIDE TO MOST LIFE-FORMS, YOU CAN HARDLY CALL IT LIVING.

HOLD ON, ALBERT, ISN'T THE ROTTEN EGG CHEMICAL ALSO IN STINK BOMBS? IT CAN'T BE THAT POISONOUS?

WELL, IT ONLY TAKES A MINISCULE AMOUNT TO MAKE A GOOD STINK. EVEN 600 MOLECULES OF HYDROGEN SULFIDE FOR EVERY MILLION MOLECULES OF AIR COULD KILL YOU, AND AT THAT CONCENTRATION, YOUR SENSE OF SMELL WOULD BE SO OVERPOWERED, YOU COULDN'T EVEN SMELL IT ANYMORE.

EVERY OTHER LIVING THING DEPENDS ON SUNLIGHT, SO THEY WOULD EVENTUALLY DIE OFF, IF THEY DIDN'T FREEZE TO DEATH FIRST.

BUT HUMANS DON'T NEED SUNLIGHT TO EAT.

NO, BUT WE EAT PLANTS, AND THEY MAKE THEIR OWN FOOD OUT OF SUNLIGHT.

HAMBURGERS AREN'T PLANTS!

BUT COWS LIVE OFF GRASS AND BURGER BUNS ARE MOSTLY MADE FROM WHEAT.

EVEN CARNIVORES, LIKE LIONS, GET THEIR FOOD FROM PLANTS AT SOME POINT DOWN THE FOOD CHAIN. INDIRECTLY, WE ARE ALL VEGETARIANS.

HOW ABOUT THAT MONSTER UNDERGROUND FUNGUS YOU TOLD ME ABOUT? THAT CAN'T NEED LIGHT.

SOME BACTERIA AND FUNGI DO ACTUALLY GLOW WHILE THEY ARE FEEDING OFF ROTTING PLANTS OR ANIMALS, BUT, SADLY, WE DON'T. MIND YOU, IT MIGHT MAKE KIDS EAT MORE VEGETABLES IF WE DID GLOW IN THE DARK AFTERWARD!

WHAT ABOUT FIREFLIES? THEY GLOW, DON'T THEY?

YES, BUT ONLY WHEN THEY'RE MATING. FIREFLIES MATE AT NIGHT AND USE FLASHES OF LIGHT TO ATTRACT ONE ANOTHER. EACH SPECIES OF FIREFLY HAS ITS OWN CODE OF FLASHES, SO THEY CAN AVOID AN EMBARRASSING MEETING IN THE DARK WITH THE WRONG KIND OF FIREFLY.

IS LIGHT IMPORTANT FOR REPRODUCTION, TOO?

OF COURSE. THE BEAUTIFUL PLUMAGE OF BIRDS IS ONLY THERE TO BE SEEN BY POTENTIAL MATES.

MOST HUMANS CHOOSE CLOTHES BECAUSE OF HOW THEY LOOK IN THEM. IF THERE WAS NO LIGHT, EVOLUTION WOULDN'T HAVE GENERATED EYES...

HMM... MAYBE I'LL JUST CUT THE LIGHTS

...AND NO ONE WOULD CARE WHAT THEY LOOKED LIKE. SO LIGHT DRIVES THE ENTIRE FASHION INDUSTRY.

LIGHT

BUT SOME ANIMALS LIVE WHERE THERE IS NO LIGHT OR VERY LITTLE LIGHT, AND HAVE LITTLE VISION. HAVE YOU HEARD THE OLD SAYING, "BEAUTY IS IN THE EYE OF THE BEHOLDER"? WELL, IF THERE'S NO ONE TO DO THE BEHOLDING, THEN WHY SHOULD EVOLUTION BOTHER TO MAKE YOU BEAUTIFUL? THIS PROBABLY EXPLAINS WHY MOLES, BATS, AND DEEP SEA FISH AREN'T SO ATTRACTIVE.

BUT APART FROM THESE CREATURES THAT LIVE IN VERY DARK PLACES, NEARLY ALL ANIMALS, EVEN THE VERY SIMPLEST, CAN REACT TO LIGHT. NOT ALL CAN "SEE" AS YOU MIGHT UNDERSTAND IT, BUT EVEN TINY MICROSCOPIC ORGANISMS CAN DETECT LIGHT AND MOVE TOWARD OR AWAY FROM IT.

SO WHAT DO YOU NEED TO BE ABLE TO SEE?

AN EYE, A BRAIN, AND A BIT OF LIGHT.

A COMPLEX EYE IS USELESS IF IT'S ATTACHED TO A BRAIN THE SIZE OF A PINHEAD. AND ANY EYE IS USELESS IF YOU LIVE SOMEWHERE WITHOUT LIGHT.

CHAPTER 30: ARE WE ALONE? LISTENING FOR EXTRATERRESTRIAL LIFE

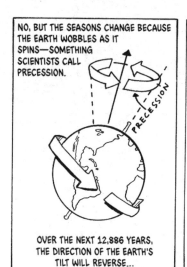

MIND YOU, FUEL FROM PLANTS ISN'T A NEW IDEA. WHEN THE DIESEL ENGINE WAS FIRST INVENTED, IT WAS DESIGNED TO USE PEANUT OIL, AND WHEN HENRY FORD DESIGNED THE MODEL T, HE PLANNED TO USE ALCOHOL, OR ETHANOL, AS FUEL.

WOW, SO HE KNEW ABOUT CLIMATE CHANGE EVEN THEN?

NOT AT ALL! HE WAS A BUSINESS MAN AND THOUGHT IT WOULD BE CHEAPER.

SO WHY AREN'T WE USING ALCOHOL IN OUR CARS TODAY?

OH, THE USUAL REASON: POLITICS. IN 1919, ALCOHOL WAS BANNED IN THE UNITED STATES DURING THE PROHIBITION ERA, AND POWERFUL OIL INTERESTS PUSHED FOR GASOLINE TO BE USED INSTEAD.

IMAGINE HOW DIFFERENT THE WORLD WOULD BE IF CARS HAD BEEN BURNING ETHANOL FOR 90 YEARS INSTEAD OF FOSSIL FUELS—NOT ONLY WOULD THE MIDDLE EAST BE A DIFFERENT PLACE BUT ALSO WORLD POLITICS. BETTER? WHO KNOWS? BUT CERTAINLY THERE'D BE LESS CARBON DIOXIDE IN THE ATMOSPHERE.

SO ARE HUMANS REALLY RESPONSIBLE FOR THE INCREASE IN TEMPERATURE?

WELL, THE EARTH'S TEMPERATURE HAS FLUCTUATED FOR HUNDREDS OF MILLIONS OF YEARS.

IT'S ONLY BEEN 10,000 YEARS SINCE THE LAST ICE AGE.

LOTS OF THINGS APART FROM CARBON DIOXIDE MIGHT HAVE CAUSED THAT, INCLUDING CHANGES IN THE EARTH'S ORBIT...

...TINY VARIATIONS IN HOW BRIGHTLY THE SUN SHINED...

...MASSIVE VOLCANOES...

...AND EVEN THE EARTH BEING HIT BY AN ASTEROID.

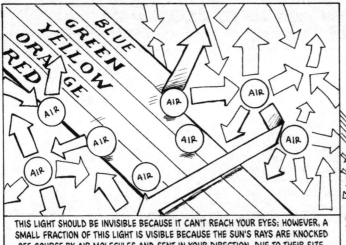
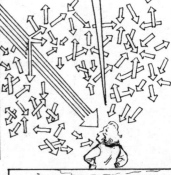

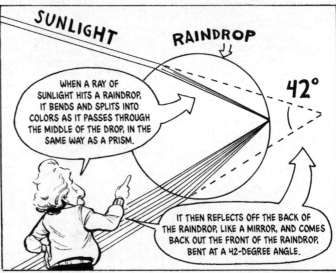

CHAPTER 33:
MORE BLACK HOLES AND WHY THE SPEED OF LIGHT ISN'T CONSTANT AFTER ALL

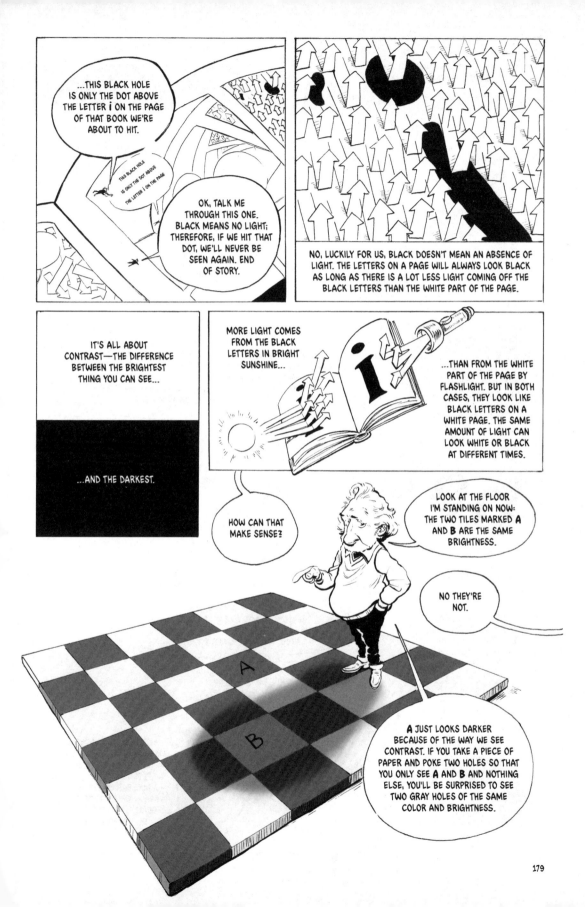

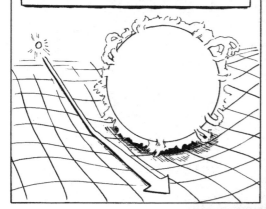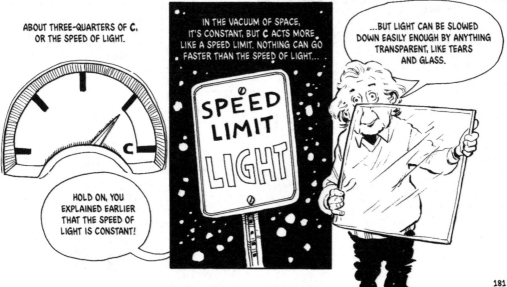

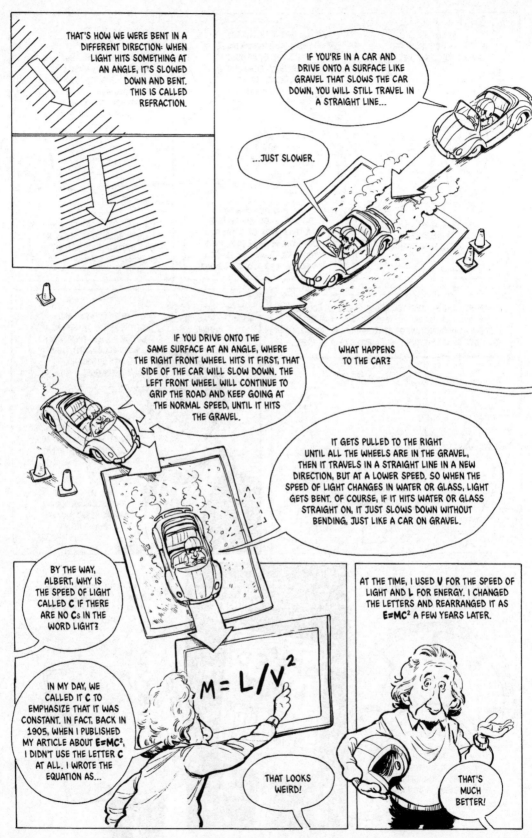

CHAPTER 34: THE CURIOUS SENSATION OF BEING SEEN

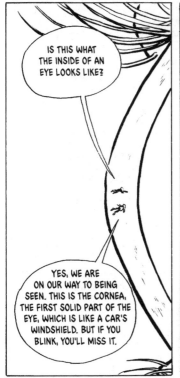

IS THIS WHAT THE INSIDE OF AN EYE LOOKS LIKE?

YES, WE ARE ON OUR WAY TO BEING SEEN. THIS IS THE CORNEA, THE FIRST SOLID PART OF THE EYE, WHICH IS LIKE A CAR'S WINDSHIELD. BUT IF YOU BLINK, YOU'LL MISS IT.

HUH?

YOU'VE MISSED IT! WE ONLY SPENT 2 TRILLIONTHS OF A SECOND GOING THROUGH THE CORNEA, WHICH IS JUST 2 HUNDREDTHS OF AN INCH (0.5 MM) THICK. NOW WE'VE GOT AN EIGHTH OF AN INCH (3 MM) OF SALTY WATER TO GET THROUGH, THEN ANOTHER BLACK HOLE.

A REAL ONE THIS TIME?

NO, THIS BLACK HOLE IS THE PUPIL OF THE EYE. WE'LL SAIL STRAIGHT THROUGH AND OUT THE OTHER SIDE.

IF PUPILS DON'T DO ANYTHING, WHY DO OUR EYES HAVE THEM?

TO CONTROL THE AMOUNT OF LIGHT ENTERING THE EYE. REMEMBER HOW I INVENTED A TYPE OF REFRIGERATOR? WELL, IN 1936, I INVENTED A CAMERA, TOO. LIKE A PUPIL, IT AUTOMATICALLY CONTROLLED HOW MUCH LIGHT REACHED THE FILM.

BUT ALL CAMERAS DO THAT!

THEY MIGHT NOW, BUT THEY DIDN'T BACK THEN. EVERYTHING HAS TO BE INVENTED BY SOMEONE!

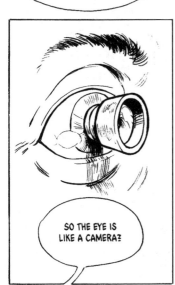

SO THE EYE IS LIKE A CAMERA?

THAT'S RIGHT, EYES AREN'T MADE OF GLASS, BUT THEY BEND LIGHT IN THE SAME WAY. WE'VE ALREADY TALKED ABOUT JOHANNES KEPLER, THE MAN WHO FIGURED THAT OUT.

DIDN'T HE ALSO WORK OUT HOW THE PLANETS MOVE?

EXACTLY, IN ADDITION TO HIS THREE LAWS OF PLANETARY MOTION, KEPLER INVENTED THE MODERN SCIENCE OF OPTICS, IN 1604.

ASTRONOMIA PARS OPTICA

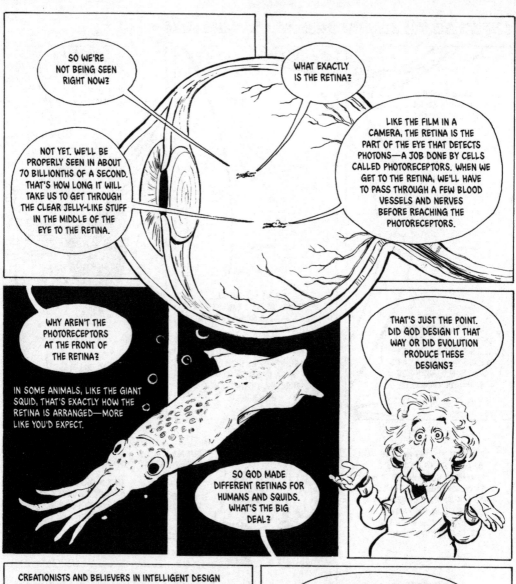

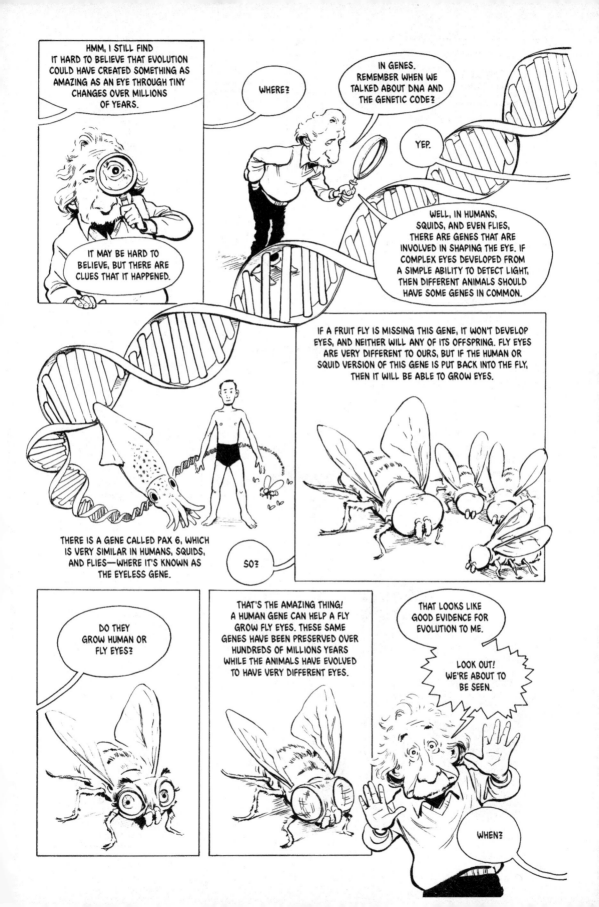

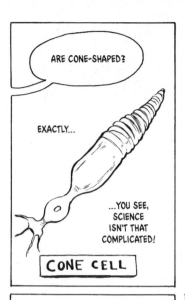

CHAPTER 35:
WHAT HAPPENS IN YOUR BRAIN AND WHAT HAPPENED TO ALBERT EINSTEIN'S BRAIN

SO HERE WE ARE INSIDE YOUR BRAIN.

MY BRAIN?

SORRY, I'M TALKING TO THE FOLKS OUT THERE. HEY, YOU, READING THIS BOOK! WE'RE INSIDE YOUR HEAD NOW.

WANT TO KNOW WHERE? PLACE YOUR FINGERS ON TOP OF YOUR SKULL. NOW RUN THEM STRAIGHT BACK, FEELING THE LUMPS AND BUMPS AS YOU GO.

AT THE BACK OF YOUR SKULL, NEAR THE TOP OF YOUR NECK, IS A BONY LUMP CALLED THE INION.

GO ON, FEEL WHERE IT IS. GOT IT?

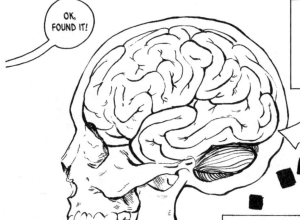

OK, FOUND IT!

GOOD, NOW, UNDERNEATH THIS BIT OF BONE, THERE'S A THIN, PINK, WRINKLY LAYER ONLY AN EIGHTH OF AN INCH (3 MM) THICK. THIS IS YOUR PRIMARY VISUAL CORTEX, WHERE BILLIONS UPON BILLIONS OF NEURONS ARE TRYING TO MAKE SOME SENSE OF WHAT YOU'RE SEEING, LINKING THE SHAPES OF THESE BLACK MARKS ON THE PAGE INTO LETTERS, WORDS, AND IDEAS.

letters, words, ideas

THE LIGHT FROM YOUR EYES IS NOW A PATTERN OF FLICKERING ACTIVITY IN BILLIONS OF BRAIN CELLS THAT STILL HAS TO BE DECODED BEFORE WE CAN REALLY CLAIM TO HAVE BEEN SEEN. THIS IS PROBABLY ONE OF THE MOST DIFFICULT THINGS YOUR BRAIN DOES.

WHAT'S SO DIFFICULT ABOUT SEEING?

COMPUTERS CAN CALCULATE MILLIONS OF TIMES FASTER THAN HUMANS AND CAN BEAT ALMOST ANYONE IN A GAME OF CHESS, BUT THEY AREN'T GOOD AT SEEING.

SEEING DOESN'T FEEL HARD TO DO. I CERTAINLY FIND IT EASIER THAN PLAYING CHESS.

THAT'S BECAUSE HUMAN BRAINS ARE BUILT FOR SEEING, NOT PLAYING CHESS.

ALMOST HALF YOUR BRAIN IS WORKING ON VISION IN ONE WAY OR ANOTHER.

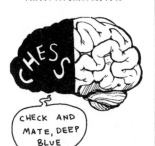

IF HALF YOUR BRAIN WAS DEDICATED SOLELY TO PLAYING CHESS, THEN CHESS WOULD BE PRETTY EFFORTLESS, TOO!

CHECK AND MATE, DEEP BLUE

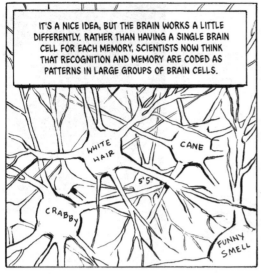

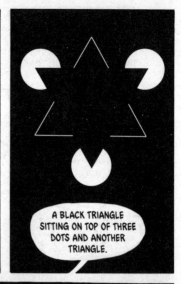

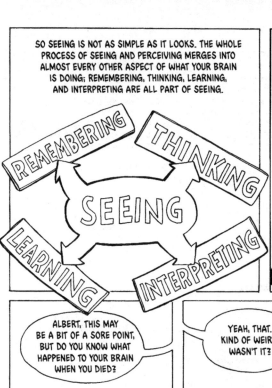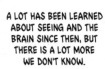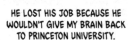

ERWIN SCHRÖDINGER, WHO MADE THE BREAKTHROUGH IN QUANTUM MECHANICS, WENT ON TO WRITE A LITTLE BOOK IN 1944 CALLED *WHAT IS LIFE?*, BASED ON THREE LECTURES HE GAVE AT TRINITY COLLEGE DUBLIN IN EARLY 1943.

IN THIS BOOK, HE PREDICTED THAT LIFE NEEDED SOME GENETIC CODE IN THE FORM OF WHAT HE CALLED AN APERIODIC CRYSTAL. JAMES WATSON WENT ON TO READ THIS BOOK, WHICH SET HIM ON THE PATH TO DISCOVER THE STRUCTURE OF DNA WITH FRANCIS CRICK, IN 1953.

SO HUMAN UNDERSTANDING HAS TRAVELED FAR FASTER THAN LIGHT EVER COULD—IT'S THE ONE THING IN THE UNIVERSE THAT DOES BREAK EINSTEIN'S RULE ABOUT NOTHING GOING FASTER THAN THE SPEED OF LIGHT, APART FROM IMAGINATION, OF COURSE. THE BIG QUESTION THAT NO ONE CAN ANSWER IS, WHY ARE ALL THESE LAWS HERE IN THE FIRST PLACE?

SO DID GOD INVENT THE RULES AND THEN JUST SIT BACK AND LET THE UNIVERSE UNFOLD FOR THE NEXT 14 BILLION YEARS?

SOME OF THE LAWS SEEM SO SIMPLE AND ELEGANT THAT IT'S HARD TO IMAGINE THEY'RE JUST THE RANDOM RESULTS OF A HUGE COSMIC ACCIDENT. TO MATHEMATICIANS AND PHYSICISTS, THESE EQUATIONS EVEN APPEAR TO BE BEAUTIFUL.

THE SCIENCE FICTION WRITER ARTHUR C. CLARKE, WHO WROTE *2001: A SPACE ODYSSEY*, INVENTED THREE LAWS ABOUT PROGRESS...

FIRST LAW: "WHEN A DISTINGUISHED BUT ELDERLY SCIENTIST STATES THAT SOMETHING IS POSSIBLE, HE IS ALMOST CERTAINLY RIGHT. WHEN HE STATES THAT SOMETHING IS IMPOSSIBLE, HE IS VERY PROBABLY WRONG."

SECOND LAW: "THE ONLY WAY OF DISCOVERING THE LIMITS OF THE POSSIBLE IS TO VENTURE A LITTLE WAY PAST THEM INTO THE IMPOSSIBLE."

THIRD LAW: "ANY SUFFICIENTLY ADVANCED TECHNOLOGY IS INDISTINGUISHABLE FROM MAGIC."

IMAGINE HOW THE THINGS YOU TAKE FOR GRANTED TODAY WOULD LOOK TO LORD KELVIN IF HE WAS TAKEN ON A 100-YEAR JOURNEY INTO HIS FUTURE. SUPERSONIC AIRPLANES, SPACE TRAVEL, MICROWAVE OVENS, AND COMPUTERS WOULD ALL LOOK LIKE SOME FORM OF MAGIC.

WHAT SEEMS LIKE SCIENCE FICTION NOW COULD, WITH THE HELP OF THE NEXT GENERATION OF SCIENTISTS, BE JUST AS REAL AS ALL OF THESE THINGS. SOMETIMES IT TAKES A LEAP OF IMAGINATION TO START BELIEVING THAT THERE ARE THINGS STILL TO BE DISCOVERED.

MAYBE SOMEONE READING THIS WILL GO ON TO PROVE THE IMPOSSIBLE REALLY IS POSSIBLE.

INDEX

A

A, OR ADENOSINE, 154
ADAMS, DOUGLAS, 176
ADENOSINE, OR A, 154
ADENOSINE TRIPHOSPHATE, OR ATP, 160
ALCHEMY, 52-53
ALICE'S ADVENTURES IN WONDERLAND, 54, 97
ALIENS, 63, 111, 115-116, 132-133, 163-167
ALRAI, OR GAMMA CEPHEI, 130
AMINO ACID, 155
AMOEBA DUBIA, 155
ANAXAGORAS OF CLAZOMENAE, 21
ANTIMATTER, 39-40
ANTIMATTER VS MATTER, 39-40
ARABIC SCHOLARS, ANCIENT, 19-20, 68
ARCHIMEDES, 138
ARCHIMEDES' PRINCIPLE, 138
ARIEL, 136
ARISTARCHUS OF SAMOS, 120
ARMILLARIA OSTOYAE, OR HONEY MUSHROOM, 156, 160
ASTRONOMY, 17
ATOM, 24-29, 53, 62, 78, 84-85, 102-103, 154
ATOMIC BOMB, OR HYDROGEN BOMB, 23, 28-32, 58-59
ATP, OR ADENOSINE TRIPHOSPHATE, 160

B

BACON, FRANCIS, 70
BARYOGENESIS THEORY, 40
BELL, JOCELYN, 63
BELL LABS, 45
BENEDETTI, LAURA, 136
BETELGEUSE, 20
BIBLE, EVENTS OF, 21-22, 33
BIG BANG, 33-37, 39-43, 93, 128
BIG CRUNCH, 43
BIOFUEL, 170-171
BLACK HOLE, 63-66, 106
BLACK STAR, 55
BOHR, NIELS, 81-82, 84-85
BRAGG, WILLIAM, 77
BRAHE, TYCHO, 123-124
BRAIN, 188-194, 201
BROWN DWARF, 60

C

C, OR CYTOSINE, 154
C, OR SPEED OF LIGHT, 29, 38, 86-90, 95-104, 108-111, 181-182
CAMBRIDGE UNIVERSITY, 46, 52, 63
CAMERA, INVENTION OF, 183
CAMPBELL, BRUCE, 130
CAR AIRBAG, INVENTION OF, 39
CARROLL, LEWIS, 195, 200
CARTER, JIMMY, 115-116
CAT FLAP, INVENTION OF, 85
CATHOLIC CHURCH, 37, 121-122
CAVENDISH, HENRY, 54-56
CEPHEUS, 130
CHARON, 118
CHEMICAL BOND, 26
CHEMICAL REACTION, 28, 157
CHINESE ASTRONOMERS, ANCIENT, 58
CHLOROPHYLL, 160
CLARKE, ARTHUR C., 199
CLIMATE CHANGE, OR GLOBAL WARMING, 141, 169-172
CODON, 154-155
COMET, 113-114
CONE, 186-187
CONSTELLATION, 19
CONSTRUCTIVE INTERFERENCE, 74, 79-80
CONTRAST, 179
COPERNICUS, OR KOPPERNIGK, NIKLAS, 121-122
CORDOBA, SPAIN, 68
CORNEA, 183
CORNELL UNIVERSITY, 194
CORONA, 150
CORPUSCLE, 72
COSMIC EGG, 42
COSMIC RAY, 102-103
COSMOLOGICAL CONSTANT, 36, 44
CREATION, 21-22
CRICK, FRANCIS, 157-158, 197
CRUSADES, 68
CYGNUS, OR THE SWAN, 19-20, 66, 131
CYTOSINE, OR C, 154

D

DA VINCI, LEONARDO, 69
DARK ENERGY, 44
DARK MATTER, 43-46
DARWIN, CHARLES, 22, 156
DE LAPLACE, PIERRE SIMON, 65
DE MAILLET, BENOÎT, 157
DE REVOLUTIONIBUS ORBIUM COELESTIUM, OR *ON THE REVOLUTION OF THE CELESTIAL SPHERES*, 121-122
DELFOSSE, XAVIER, 133
DEMOCRITUS OF ABDERA, 24
DENEB, 17-20
DEOXYRIBONUCLEIC ACID, OR DNA, 154-157, 185, 196
DIALOGUE CONCERNING THE TWO CHIEF WORLD SYSTEMS, 122
DINOSAUR FOSSILS, 22
DIRECTED PANSPERMIA, 157
DNA, OR DEOXYRIBONUCLEIC ACID, 154-157, 185, 196
DOPPLER, CHRISTIAN, 38
DOPPLER EFFECT, 38
DOUBLE HELIX, 154
DRAKE, DR. FRANK, 164-167
DRAKE EQUATION, 164-167
DWARF PLANET, 117-118
DYSNOMIA, 117

E

E, OR ENERGY, 28-30, 78, 180
E=HF, 78-79
$E=mc^2$, 29-32, 39, 96, 180, 182, 196
EARTH, 21-22, 32, 55-56, 125, 136, 145-146, 153, 163, 168-175
EDDINGTON, ARTHUR, 106, 151-152
EGYPTIAN ASTRONOMERS, ANCIENT, 123
EINSTEIN, ALBERT, 14, 29-32, 34-36, 39, 44, 57, 77-79, 85, 87, 91-106, 109-111, 146-152, 181, 193, 196, 200
ELECTROMAGNETIC SPECTRUM, 75-76, 87, 196
ELECTRON, 25-27, 62, 78
ELECTRON DEGENERACY, 62
ELEMENT, 24-26, 61
ENERGY, OR E, 28-30, 78, 180
ERIS, 117
ETHER, 89-90, 101
"EUREKA," 37, 138
EVERYTHING, THEORY OF, 85
EVOLUTION, 22, 156-157, 184-185
EVOLUTION VS INTELLIGENT DESIGN, 184
EYE, 161-162, 181, 183-190

F

F, OR FREQUENCY, 75, 78
FAITH VS SCIENCE, 23
FITZGERALD, GEORGE FRANCIS, 101
FITZGERALD CONTRACTION, 101
581G, 133
FLAMEL, NICHOLAS, 53
FLYER 1, 198
FORD, HENRY, 171
FOSSIL FUEL, 169-172
FOUCAULT, LÉON, 89
FRAME OF REFERENCE, 92-93
FREQUENCY, OR F, 75, 78
FRESNEL, AUGUSTIN, 72
FRIEDMANN, ALEXANDER ALEXANDROVICH, 34-36
FUSIFORM GYRUS, 190

G

G, OR GUANINE, 154
GALAPAGOS ISLANDS, 156
GALILEO, 69, 87-89, 122-123, 125-127
GAMMA CEPHAI, OR ALRAI, 130
GEODESIC, 148-149, 152
GENE, 155, 185
GENERAL RELATIVITY, 91, 104-106, 147-152, 181
GENETIC CODE, 155-156, 185
GLIESE 581, 133
GLIESE 581C, OR YMIR, 133
GLOBAL WARMING, OR CLIMATE CHANGE, 141, 169-172
GLUON, 27
GODFATHER, THE, 118-119
"GOLDILOCKS AND THE THREE BEARS," 132
GOLDILOCKS ZONE, 132-133, 135, 153, 165
GOSSE, PHILIP, 22
GRANDMOTHER CELL, 190
GRAVITY, 43-44, 47-49, 51, 54-56, 58-66, 104-106, 112-114, 129-130, 144-152
GRAVITY, LAW OF, 48-49, 51, 54-56, 113-114, 146
GREEKS, ANCIENT, 11, 16, 20-21, 24-25, 118-120, 138
GREENHOUSE GAS, 141, 145, 169-170
GUANINE, OR G, 154

H

H, OR PLANCK'S CONSTANT, 78
HALL, DAVID B., 103
HALLEY, EDMUND, 113-114
HALLEY'S COMET, 113
HARRY POTTER AND THE PHILOSOPHER'S STONE, 53
HARVEY, DR. THOMAS, 193
HEAT WAVES, OR INFRARED RADIATION, 141
HEISENBERG, WERNER, 80
HESCHEL, RABBI ABRAHAM JOSHUA, 177
HETRICK, JOHN W., 39
HEWISH, ANTONY, 63
HINDU MATHEMATICIANS, ANCIENT, 20
HIROSHIMA, JAPAN, 31
HITCHHIKER'S GUIDE TO THE GALAXY, THE, 176
HONEY MUSHROOM, OR *ARMILLARIA OSTOYAE*, 156, 160
HOYLE, FRED, 37, 157
HUBBLE, EDWIN, 37-38
HUBBLE SPACE TELESCOPE, 128
HUNTER, THE, OR ORION, 20
HUYGENS, CHRISTIAAN, 72
HYDRA, 118
HYDROGEN BOMB, OR ATOMIC BOMB, 23, 28-32, 58-59

I

IAU, OR INTERNATIONAL ASTRONOMICAL UNION, 117-118, 134-135
INDIAN ASTRONOMERS, ANCIENT, 120
INFRARED RADIATION, OR HEAT WAVES, 141
INION, 189
INTELLIGENT DESIGN, 184
INTELLIGENT DESIGN VS EVOLUTION, 184
INTERFERENCE, 45, 73
INTERNATIONAL ASTRONOMICAL UNION, OR IAU, 117-118, 134-135
IRVING, WASHINGTON, 181

J

JANSKY, KARL, 45
JEANLOZ, RAYMOND, 136
JUPITER, 88, 118-120, 126-127, 138-140

K

KANIZSA TRIANGLE, 192
KELVIN, LORD, 198
KEPLER, JOHANNES, 124, 127, 130-131, 183, 187, 196
KEPLER TELESCOPE, 131
KING HIERON II OF SYRACUSE, 138
KOPPERNIGK, NIKLAS, OR COPERNICUS, 121-122
KREBS, NICKOLAUS, 120

L

LANDSKRONA CASTLE, 124
LAST SUPPER, THE, 69
LAW OF GRAVITY, 48-49, 51, 54-56, 113, 146

LAWS OF MOTION, 50–51, 55
LAWS OF PLANETARY MOTION, 127
LAWS OF PROBABILITY, 80, 82–83
LAWS OF THERMODYNAMICS, 180, 186
LEANING TOWER OF PISA, 168
LEMAÎTRE, GEORGES-HENRI, 35–36, 42–43
LETTVIN, JERRY, 190
LEVIN, GILBERT, 144
LEWIS, GILBERT, 78
LGM-1, 63
LIFE, MEANING OF, 176–177
LIFE, ORIGINS OF, 156–158
LIGHT, 41–42, 71–80, 90, 106, 147–152, 159–162, 173–176, 179–188, 196, 201
LIGHT, SPEED OF, 29, 38, 86–90, 95–104, 108–111, 181–182
LIGHT, THEORY OF, 72, 173–176
LIGHT-SECOND, 15
LIGHT-YEAR, 15
LIGHT CONE, 110–111
LIGHT WAVE, 41–42, 72–76, 79–80, 90
LINCOLNSHIRE, ENGLAND, 47–48
LOWELL, PERCIVAL, 143–144

M

M, OR MASS, 29–30
MACHO, OR MASSIVE COMPACT HALO OBJECT, 46
MAD HATTER, 54
MAGNETRON, 42
MARS, 141–145, 169
MASS, OR M, 29–30
MASS VS WEIGHT, 55
MASSIVE COMPACT HALO OBJECT, OR MACHO, 46
MASTOID PROCESS, 190
MATRIX, THE, 198
MATTER, 39–40, 43, 46
MATTER VS ANTIMATTER, 39–40
MEANING OF LIFE, 176–177
MENDELEEV, DMITRI IVANOVICH, 24
MERCURY (ELEMENT), 53–54
MERCURY (PLANET), 120, 146
METABOLISM, 160
METHANE, 134, 136
MICHELL, REVEREND JOHN, 54
MICHELSON, ALBERT, 90
MICROWAVE, 40–42
MICROWAVE OVEN, 14, 41
MILKY WAY GALAXY, 15, 58, 64, 107
MILLER, STANLEY, 158
MINKOWSKI, HERMANN, 102
MIRANDA, 136
MODEL T, 171
MOLECULE, 26, 154, 156
MOLYBDENUM, 157–158
MOLYNEUX, WILLIAM, 193
MONA LISA, 69
MOON (EARTH), 125, 136, 145–146, 163
MOORS, 68
MORLEY, EDWARD, 90
MOTION, LAWS OF, 50–51, 55
MUON, 102–104

N

NAGASAKI, JAPAN, 31
NASA, 114–116, 128, 131, 135, 143–144, 163
NATURAL PHILOSOPHER, 69
NEPTUNE, 118–119, 134–136
NEURON, 189

NEUTRON, 26–27, 62
NEUTRON STAR, 62–64
NEW YORK TIMES, THE, 151
NEWTON, HANNAH, 48, 52
NEWTON, ISAAC, 47–52, 54–57, 65, 68, 71–72, 85, 102, 105, 113, 127–128, 173–176, 196
NEWTONIAN MECHANICS, 51
NONEXISTANT PAST, 22
NUCLEAR FISSION, 28
NUCLEAR FUSION, 28, 30
NUCLEUS, 25–26

O

OBERON, 136
OF THE PROFICIENCE AND ADVANCEMENT OF LEARNING, DIVINE AND HUMAN, 70
OLYMPIANS, 119
OMPHALOS, 22
ON THE REVOLUTION OF THE CELESTIAL SPHERES, OR *DE REVOLUTIONIBUS ORBIUM COELESTIUM*, 121–122
OORT, JAN, 112, 114
OORT CLOUD, 112–114
OPSIN, 186
OPTIC NERVE, 188
OPTICS, MODERN SCIENCE OF, 183
ORGEL, LESLIE, 157
ORIGIN OF SPECIES, THE, 22
ORIGINS OF LIFE, 156–158
ORION, OR THE HUNTER, 20

P

P CYGNI, 18–19
PARADOX, 81
PARALLEL UNIVERSE, 40, 82
PARSBERG, MANDERUP, 124
PAX 6, 185
PERIODIC TABLE, 24, 26
PHILOSOPHER'S STONE, 153
PHOTOELECTRIC EFFECT, 77–78
PHOTON, 34, 78, 180, 186
PHOTORECEPTOR, 184, 186–188
PHOTOSYNTHESIS, 160, 170
PLAGUE OF 1665, 48
PLANCK, MAX, 78, 102
PLANCK'S CONSTANT, OR H, 78
PLANET, 19, 117–124, 126–127, 129–132, 134–136, 165
PLANETARY MOTION, LAWS OF, 127
PLANTS, 159–160, 170–171
PLATO, 117–118
PLUTO, 117–119, 134
PLUTOID, 117–118
POE, EDGAR ALLAN, 37
POLE STAR, 130
PONTIFICAL ACADEMY OF SCIENCES, 37
POPE PIUS XI, 37
POPE PIUS XII, 37
POTTER, HARRY, 53
PRECESSION, 169
PRIMARY VISUAL CORTEX, 189
PRIMORDIAL PARTICLE, 37
PRINCETON UNIVERSITY, 193
PRINCIPE ISLAND, 106, 151
PRISM, 71–72, 173, 176
PROBABILITY, LAWS OF, 80, 82–83
PROHIBITION, 171
PROTEIN, 155
PROTON, 26–27, 62
PROXIMA CENTAURI, 18, 115
PULSAR, 63
PUPIL, 183

Q

QUANTA, 77–78
QUANTUM MECHANICS, 77, 79–85
QUARK, 27

R

RADIO ASTRONOMY, 45
RADIO TELESCOPE, 45, 63
RADIO WAVE, 41–42, 45, 63, 87
RAINBOW, 71–72, 173–177
RAYLEIGH SCATTERING, 173–174
RED DWARF, 60
RED GIANT, 59
RED SHIFT, 38
RED SPOT, 139
REFLECTING TELESCOPE, 128
REFRACTION, 182
REFRIGERATOR, INVENTION OF, 14
RELATIVITY, 69, 91–92
RELATIVITY, THEORY OF, 87, 91–106, 109–110, 146–152, 181, 196
RENAISSANCE, 69
RETINA, 184, 186, 190
ROCKY IV, 141
ROD, 186
ROMANS, ANCIENT, 19–20, 118–120
RØMER, OLE, 88
ROOSEVELT, FRANKLIN D., 31–32
ROSSI, BRUNO, 103
ROSWELL, NEW MEXICO, 111
RULLOFF, EDWARD, 194

S

SATURN, 118–120, 136–138
SCHRÖDINGER, ERWIN, 83–85, 197
SCHRÖDINGER'S CAT, 83–85
SCHWARZSCHILD, KARL, 64–65
SCIENCE VS FAITH, 23
SCIENTIFIC PROCESS, 70
SEARCH FOR EXTRA-TERRESTRIAL INTELLIGENCE, OR SETI, 164–167
SETI, OR SEARCH FOR EXTRA-TERRESTRIAL INTELLIGENCE, 164–167
SMITH, REVEREND BARNABAS, 52
SOBRAL, BRAZIL, 151
SOLAR ECLIPSE, 106, 149–152, 178, 181
SOLAR SYSTEM, 120–127, 129
SOUND, SPEED OF, 18, 38, 87
SOUND WAVE, 34
SPACE, 107
SPACE-TIME, THEORY OF, 102
SPECIAL RELATIVITY, 91–104
SPEED OF LIGHT, OR C, 29, 38, 86–90, 95–104, 108–111, 181–182
SPEED OF SOUND, 18, 38, 87
SPENCER, PERCY, 42
STAR, 16–23, 28–30, 32, 57–66, 106–108, 120, 126, 129–132, 140, 164–165, 178
STAR WARS, 43
STEREOPSIS, OR 3-D VISION, 191
STROMATOLITE, 156
STYX, 118
SUBATOMIC PARTICLE, 46
SUN, 18, 21–23, 113–114, 120–122, 125, 140, 147–150
SUPERGIANT, 18
SUPERNOVA, 57–58, 61–62
SWAN, THE, OR CYGNUS, 19–20, 66, 131
SZILARD, LEO, 14, 31

T

T, OR THYMIDINE, 154
TAYLOR, GEOFFREY, 79–80
TELESCOPE, 45, 63, 66, 125, 128, 131, 196
TEMPORAL LOBE, 190
TERRAFORMING, 142
THEORY OF EVERYTHING, 85
THEORY OF LIGHT, 72, 173–176
THEORY OF RELATIVITY, 87, 91–106, 109–110, 146–152, 181, 196
THEORY OF SPACE-TIME, 102
THERMODYNAMICS, LAWS OF, 180, 186
THREE LAWS OF PROGRESS, 199
3-D VISION, OR STEREOPSIS, 191
THROUGH THE LOOKING GLASS, 80, 200
THYMIDINE, OR T, 154
TIME, 33, 57, 93–104, 106
TIMES OF LONDON, THE, 151
TITANIA, 136
TITANS, 118–119
TRITON, 135
TROJAN WAR, 11, 16
TROPOSPHERE, 173
TRUMAN, HARRY S., 32
2001: A SPACE ODYSSEY, 199

U

UMBRIEL, 136
UNCERTAINTY PRINCIPLE, 80
UNIVERSE, 34–44, 93
URANUS, 118–119, 136
UREY, HAROLD, 158
USSHER, JAMES, 21

V

VENERA SPACE PROBES, 146
VENUS, 120, 145–146, 169
VIKING SPACE PROBES, 143–144
VISION, 161–162, 188–193, 201
VON MAYER, JULIUS ROBERT, 21
VOYAGER 1, 114–116
VOYAGER 2, 115–116
VOYAGER MESSAGE, 115–116

W

WALKER, G. A. H., 130
WATSON, JAMES, 196–197
WAVE FUNCTION, 81–82
WAVELENGTH, 74–76
WEAKLY INTERACTING MASSIVE PARTICLE, OR WIMP, 45–46
WEIGHT VS MASS, 55
WHAT IS LIFE?, 197
WHITE DWARF, 59–60, 62
WIMP, OR WEAKLY INTERACTING MASSIVE PARTICLE, 45–46
WORLD WAR I, 150
WORLD WAR II, 31–32
WRIGHT BROTHERS, 198

X

X-1, 66
X-RAY, 66
X-RAY TELESCOPE, 66

Y

YANG, S., 130
YOUNG, THOMAS, 73–74, 76–77, 79, 187
YMIR, OR GLIESE 581C, 133

ACKNOWLEDGMENTS

IAN:

Like Albert's journey, this project has taken a while to come to fruition. Along the way, I have been helped by all the great feedback and comments from readers of my Journey by Starlight blog. I would particularly like to thank my agent, Brandi Bowles (Foundry Literary + Media), for making contact and convincing me that Journey by Starlight would make a great graphic novel.

As my drawing ability is at the level of stick-men, this seemed daunting until Brandi tracked down the brilliant Britt Spencer, who brought the story to life graphically. His dedication in completing this book made him soldier on through the pain of a broken hand. All great art is born out of suffering, and Britt's wonderful drawings are a testament to that!

On a more personal note, the person who has supported and encouraged me throughout all this is the light of my life, Jean. She has made sure I have remained sane, humble, enthused, and loved. Who could ask for more?

BRITT:

Thanks to Brandi Bowles for recognizing the potential of this book and bringing Ian and me together for this collaboration.

Thanks to Ian for writing a marvellous book and making it so easy to comprehend. It made my job infinitely easier.

Thanks to Erin Canning for putting up with the troubling schedule changes that a broken hand made necessary (this "thought experiment" of ours got a little rough at points).

Big thanks to my team of dedicated and talented interns, Allie Jachimowicz, Emily Spencer, and Lomaho Kretzmann. I'd probably still be inking if it weren't for all their help.

And lastly, thanks to a network of great friends and family that helped me get through the process without losing my mind completely.

ABOUT THE AUTHOR AND ILLUSTRATOR

DR. IAN FLITCROFT is a vision scientist and a consultant eye surgeon at the Children's University Hospital, Dublin. As a teenager, he made a close call between choosing medicine or astrophysics as a career; he hopes this book will encourage a new generation of astrophysicists, scientists, and perhaps even a few eye surgeons.

Ian studied medicine at Oxford University, where he also completed his doctorate (D.Phil) in visual physiology. He has published more than 30 scientific peer-reviewed papers and chapters in multi-author books. His writing has been featured in *Do Polar Bears Get Lonely?*, and his first novel, *The Reluctant Cannibals*, was one of the winners of the Irish Writers' Centre Novel Competition 2012 and shortlisted for Amazon's Breakthrough Novel Award in 2013.

BRITT SPENCER is an award-winning illustrator. His work has been published internationally and recognized by the distinguished New York Society of illustrators, *3x3*, and *New American Paintings*. He has illustrated three children's books and contributed hundreds of editorial spots to various publications.

Britt lives in Savannah, Georgia, where he teaches illustration at the Savannah College of Art and Design. You can see more of his work at his website, www.brittspencer.com.